Japanese Women in Science and Engineering

The gender gap in science, technology, engineering and mathematics (STEM) varies greatly from country to country, and the number of Japanese women in these fields remains relatively few. This prompts us to ask why the proportion of female scientists in Japan is still remarkably low and what measures the government, universities and research institutes are taking to address this issue.

This book sheds light on historical developments and the current gender equality situation in Japan, through the lens of women in STEM. It shows how a policy of gender equality in science and engineering has been introduced through the coordinated efforts of academia, scientific societies and the government, and how this has led to a slow but steady increase in female representation. The book draws on extensive data including interviews with government officials, scientists and educators in Japan to provide a revealing case study on how the underrepresentation of women in the fields of science, technology and engineering has been approached and dealt with by a national government. It heralds a new era for female scientists, by showcasing several programmes undertaken by government, universities and national research institutions to support multiple career paths for and the progression of female scientists in Japan.

Tracing the historical development of Japan's policies towards women in science and engineering, this book will be welcomed by students and scholars interested in Japanese studies, comparative social policy, gender studies, education policy, political economy, and the history of science and technology.

Naonori Kodate is a Lecturer in Social Policy at University College Dublin, Ireland. His main research area is comparative social policy, particularly in health care. He has co-edited a book entitled *Maternity Services and Policy in an International Context: Risk, Citizenship and Welfare Regimes* (Routledge, 2015).

Kashiko Kodate is Professor Emeritus at Japan Women's University in the Department of Mathematical and Physical Science, currently a specially appointed Professor at the University of Electro-Communications, Tokyo, and CEO of Photonic System Solutions Inc. (PSS). She was a former Director of Gender Equality Bureau at the Japan Science and Technology Agency.

Routledge Contemporary Japan Series

1 **A Japanese Company in Crisis**
Ideology, strategy, and narrative
Fiona Graham

2 **Japan's Foreign Aid**
Old continuities and new directions
Edited by David Arase

3 **Japanese Apologies for World War II**
A rhetorical study
Jane W. Yamazaki

4 **Linguistic Stereotyping and Minority Groups in Japan**
Nanette Gottlieb

5 **Shinkansen**
From bullet train to symbol of modern Japan
Christopher P. Hood

6 **Small Firms and Innovation Policy in Japan**
Edited by Cornelia Storz

7 **Cities, Autonomy and Decentralization in Japan**
Edited by Carola Hein and Philippe Pelletier

8 **The Changing Japanese Family**
Edited by Marcus Rebick and Ayumi Takenaka

9 **Adoption in Japan**
Comparing policies for children in need
Peter Hayes and Toshie Habu

10 **The Ethics of Aesthetics in Japanese Cinema and Literature**
Polygraphic desire
Nina Cornyetz

11 **Institutional and Technological Change in Japan's Economy**
Past and present
Edited by Janet Hunter and Cornelia Storz

12 **Political Reform in Japan**
Leadership looming large
Alisa Gaunder

13 **Civil Society and the Internet in Japan**
Isa Ducke

14 **Japan's Contested War Memories**
The 'memory rifts' in historical consciousness of World War II
Philip A. Seaton

15 **Japanese Love Hotels**
A cultural history
Sarah Chaplin

16 **Population Decline and Ageing in Japan – The Social Consequences**
Florian Coulmas

17 ***Zainichi* Korean Identity and Ethnicity**
David Chapman

'This book written by Professor K. Kodate, along with co-author and second son Dr N. Kodate, offers an excellent overview of historical developments in female education in Japan, while focusing on the progress made by Japanese women in STEM in recent years. Leaving aside Professor Kodate's own achievements in promoting gender equality in the field, her experience as an excellent scientist, educator, and pioneer as a woman in science who struck a great life-work balance is also reflected in the book.'

Atsuko Tōyama, *Former Minister of Education, Culture, Sports, Science and Technology*

'This book deals with the situation surrounding Japanese women in scientific research. As the authors state, "like many other countries, Japan is still in the middle of tackling the common problems known as 'glass ceiling' and 'leaky pipes' that are facing women in STEM." The book is very timely as it offers a very deep and documented study of the past and present. The analysis of the past, current laws and policies aimed at achieving equality between men and women, the important role played by scientific societies and the results in terms of statistics and surveys is a very helpful way to focus our minds on the future. I firmly believe that studies like this book will help move the situation forward.'

María J. Yzuel, *Department of Physics, Universidad Autónoma de Barcelona, Spain and Former President of SPIE, the International Society for Optics and Photonics*

'This important book explains why women's participation in STEM fields has lagged so badly in Japan relative to other industrialized nations, and describes how and why change is finally accelerating. Deeply held beliefs about women's role in Japanese society remain a serious barrier. Time does not change these – visionary and courageous leadership, laws, and data-driven remedies are required.'

Nancy Hopkins, *Amgen Inc. Professor of Biology (emerita), Massachusetts Institute of Technology, USA*

18 A Japanese Joint Venture in the Pacific
Foreign bodies in tinned tuna
Kate Barclay

19 Japanese-Russian Relations, 1907–2007
Joseph P. Ferguson

20 War Memory, Nationalism and Education in Post-War Japan, 1945–2007
The Japanese history textbook controversy and Ienaga Saburo's court challenges
Yoshiko Nozaki

21 A New Japan for the Twenty-First Century
An inside overview of current fundamental changes and problems
Edited by Rien T. Segers

22 A Life Adrift
Soeda Azembo, popular song and modern mass culture in Japan
Translated by Michael Lewis

23 The Novels of Oe Kenzaburo
Yasuko Claremont

24 Perversion in Modern Japan
Psychoanalysis, literature, culture
Edited by Nina Cornyetz and J. Keith Vincent

25 Homosexuality and Manliness in Postwar Japan
Jonathan D. Mackintosh

26 Marriage in Contemporary Japan
Yoko Tokuhiro

27 Japanese Aid and the Construction of Global Development
Inescapable solutions
Edited by David Leheny and Carol Warren

28 The Rise of Japanese NGOs
Activism from above
Kim D. Reimann

29 Postwar History Education in Japan and the Germanys
Guilty lessons
Julian Dierkes

30 Japan-Bashing
Anti-Japanism since the 1980s
Narrelle Morris

31 Legacies of the Asia-Pacific War
The Yakeato generation
Edited by Roman Rosenbaum and Yasuko Claremont

32 Challenges of Human Resource Management in Japan
Edited by Ralf Bebenroth and Toshihiro Kanai

33 Translation in Modern Japan
Edited by Indra Levy

34 Language Life in Japan
Transformations and prospects
Edited by Patrick Heinrich and Christian Galan

35 The Quest for Japan's New Constitution
An analysis of visions and constitutional reform proposals 1980–2009
Christian G. Winkler

36 Japan in the Age of Globalization
Edited by Carin Holroyd and Ken Coates

37 Social Networks and Japanese Democracy
The beneficial impact of interpersonal communication in East Asia
Ken'ichi Ikeda and Sean Richey

38 Dealing with Disaster in Japan
Responses to the Flight JL123 crash
Christopher P. Hood

39 The Ethics of Japan's Global Environmental Policy
The conflict between principles and practice
Midori Kagawa-Fox

40 **Superhuman Japan**
Knowledge, nation and culture in US–Japan relations
Marie Thorsten

41 **Nationalism, Political Realism and Democracy in Japan**
The thought of Masao Maruyama
Fumiko Sasaki

42 **Japan's Local Newspapers**
Chihōshi and revitalization journalism
Anthony S. Rausch

43 **Mental Health Care in Japan**
Edited by Ruth Taplin and Sandra J. Lawman

44 **Manga and the Representation of Japanese History**
Edited by Roman Rosenbaum

45 **Negotiating Censorship in Modern Japan**
Edited by Rachael Hutchinson

46 **EU–Japan Relations, 1970–2012**
From confrontation to global partnership
Edited by Jörn Keck, Dimitri Vanoverbeke and Franz Waldenberger

47 **Japan and the High Treason Incident**
Edited by Masako Gavin and Ben Middleton

48 **Diplomacy in Japan–EU Relations**
From the Cold War to the post-bipolar era
Oliviero Frattolillo

49 **Sound, Space and Sociality in Modern Japan**
Edited by Joseph D. Hankins and Carolyn S. Stevens

50 **Japanese Femininities**
Justin Charlebois

51 **Japan's Foreign Aid to Africa**
Angola and Mozambique within the TICAD process
Pedro Amakasu Raposo

52 **Internationalising Japan**
Discourse and practice
Edited by Jeremy Breaden, Stacey Steele and Carolyn S. Stevens

53 **Heritage Conservation and Japan's Cultural Diplomacy**
Heritage, national identity and national interest
Natsuko Akagawa

54 **Religion and Psychotherapy in Modern Japan**
Edited by Christopher Harding, Iwata Fumiaki and Yoshinaga Shin'ichi

55 **Party Politics in Japan**
Political chaos and stalemate in the 21st century
Edited by Ronald J. Hrebenar and Akira Nakamura

56 **Career Women in Contemporary Japan**
Pursuing identities, fashioning lives
Anne Stefanie Aronsson

57 **Visions of Precarity in Japanese Popular Culture and Literature**
Edited by Kristina Iwata-Weickgenannt and Roman Rosenbaum

58 **Decision-Making Reform in Japan**
The DPJ's failed attempt at a politician-led government
Karol Zakowski

59 **Examining Japan's Lost Decades**
Edited by Yoichi Funabashi and Barak Kushner

60 **Japanese Women in Science and Engineering**
History and policy change
Naonori Kodate and Kashiko Kodate

61 **Japan's Border Issues**
Pitfalls and Prospects
Akihiro Iwashita

Japanese Women in Science and Engineering

History and policy change

Naonori Kodate and Kashiko Kodate

Routledge
Taylor & Francis Group
LONDON AND NEW YORK

First published 2016 by Routledge

2 Park Square, Milton Park, Abingdon, Oxfordshire OX14 4RN
711 Third Avenue, New York, NY 10017

Routledge is an imprint of the Taylor & Francis Group, an informa business

First issued in paperback 2017

British Library Cataloguing in Publication Data
A catalogue record for this book is available from the British Library

Library of Congress Cataloging-in-Publication Data
Kodate, Naonori.
Japanese women in science and engineering : history and policy change / Naonori Kodate and Kashiko Kodate.
pages cm. – (Routledge contemporary Japan series ; 60)
Includes bibliographical references and index.
ISBN 978-1-138-81821-7 (hardback) – ISBN 978-1-315-74539-8 (ebook)
1. Women in science–Japan. 2. Women in engineering–Japan. I. Kodate, Kashiko. II. Title.
Q130.K64 2015
500.82'0952–dc23
2015001819

ISBN: 978-1-138-81821-7 (hbk)
ISBN: 978-0-8153-8390-1 (pbk)

Typeset in Times New Roman
by Cenveo Publishers Services

About the cover illustration

Cover photos: 'Average faces of women at university in Japan' (alumnae from Japan Women's University and Tokyo Woman's Christian University) created by the authors and Ms Tomoko Nakayama (photos from various years between 1906 and 2014). Images courtesy of Japan Women's University and Tokyo Woman's Christian University.

A three-dimensional wire-frame model of facial images was used to create the average face. The coordinates of the feature points and the texture information underwent averaging processing. As the individuality of the original facial images becomes negated, the common feature stands out in relief. For further details regarding how to create these average faces, see the following:

1. Harashima, H. and Naemura T. 'Face Tool, software'. http://nae-lab.org/project/face/HeikinTool/
2. Kodate, K. and Watanabe, E. (2007) 'Compact Parallel Optical Correlator for Face Recognition, and its Application, in Face Recognition', eds. Kresimir Delac and Mislav Grgic, 261–274. Vienna, Austria: I-Tech Education and Publishing.

'With nobler desires, greater earnestness and wider sympathy not limited to just a few, but taking in the many even beyond the home, the weakest of us may attain success.'

Ms Umeko Tsuda
Tsuda, U. (1984). 'The Principal's Address to the Graduate's (May 1913)',
Tsuda Umeko Bunsho, Kaiteiban
(The Writings of Umeko Tsuda: rev. edition), p. 136

'Science is a quest for understanding.'

Dame Susan Jocelyn Bell Burnell
In the interview, BBC Four, BBC 'Beautiful Minds' Jocelyn Bell Burnell on
Truth and Understanding (7 April 2010)

Contents

Figures

Tables

Acronyms and abbreviations

ARC	Administrative Reform Council
BAGES	Basic Act for Gender Equal Society
CEDAW	United Nations Committee on the Elimination of Discrimination against Women
CNRS	Le Centre national de la recherche scientifique
CSJ	Chemical Society of Japan
CSTP	Council for Science and Technology Policy
ECOSOC	United Nations Economic and Social Council
EEOL	Equal Employment Opportunity Law
EPMEWSE	Japan Inter-Society Liaison Association Committee for Promoting Equal Participation of Men and Women in Science and Engineering
ESCAP	United Nations Economic and Social Commission for Asia and the Pacific
GEM	Gender Empowerment Measure
GGI	Global Gender Gap Index
GHQ	General Headquarters, the Supreme Commander for the Allied Powers
HDI	Human Development Index
IAIs	Independent Administrative Institutions
ICWES	International Conference of Women Scientists and Engineers
ISCED	International Standard Classification of Education
JAICOWS	Japanese Association for the Improvement of Conditions of Women Scientists
JICE	Japan International Cooperation Center
JSAP	Japan Society of Applied Physics
JSPS	Japan Society for the Promotion of Science
JST	Japan Science and Technology Agency
JWEF	Japan Women Engineers Forum
JWU	Japan Women's University
Keidanren	Nihon Keizai Dantai Rengōkai (Japan Business Federation)
LDP	Liberal Democratic Party
METI	Ministry of Economy, Trade and Industry

MEXT	Ministry of Education, Culture, Sports, Science and Technology
MHLW	Ministry of Health, Labour and Welfare
MICS	Ministry of Internal Affairs and Communications
NHK	Nihon Hōsō Kyōkai (Japan Broadcasting Corporation)
NSF	National Science Foundation
OECD	Organisation for Economic Co-operation and Development
PSJ	Physical Society of Japan
RIKEN	Rikagaku Kenkyūjo (Institute of Physical and Chemical Research)
Rikejo	Rikei Joshi (science girl)
RPD	Restart Postdoctoral (Fellowship)
SCJ	Science Council of Japan
SGEPSS	Society of Geomagnetism and Earth, Planetary and Space Sciences
SJWS	Society of Japanese Women Scientists
SSHS	Super Science High School
STEM	Science, Technology, Engineering and Mathematics
TUAT	Tokyo University of Agriculture and Technology
TWCU	Tokyo Woman's Christian University
TWHNS	Tokyo Women's Higher Normal School
UN	United Nations
UNESCO	United Nations Educational, Scientific and Cultural Organization
WEF	World Economic Forum
WWII	World War Two

About the authors

Naonori Kodate (PhD in Political Science, London School of Economics and Political Science) is a Lecturer in Social Policy at the School of Applied Social Science, University College Dublin. His main research area is comparative social policy, particularly in health care. His doctoral thesis investigates the mechanisms of policy change in the hospital sector in England, Sweden and Japan. He has co-edited a book entitled *Maternity Services and Policy in an International Context: Risk, Citizenship and Welfare Regimes* (Routledge, 2015).

He was previously a researcher at King's College London (National Institute of Health Research King's Patient Safety and Service Quality Research Centre) and has published several papers (for example, in *BMJ: Quality & Safety, International Journal of Nursing Studies*) relating to patient safety and risk decision-making in various clinical settings. He is also affiliated to the University of Tokyo's Policy Alternatives Research Institute, University College Dublin Geary Institute for Public Policy and the Graduate School of Public Policy, Hokkaidō University. He has worked on projects funded by the Institute for Women's Studies, Tokyo Woman's Christian University, the Japanese Ministry of Health, Labour and Welfare, and the Japan Cancer Society.

Kashiko Kodate (DEng, University of Tokyo) is Professor Emeritus at Japan Women's University in the Department of Mathematical and Physical Science, currently a specially appointed Professor at the University of Electro-Communications, Tokyo, and CEO of Photonic System Solutions Inc. (PSS). She was a former Director of Gender Equality Bureau at the Japan Science and Technology Agency (JST, 2008–2014).

She has spent over 40 years in opto-electronics research, ranging from micro-optics to information photonics. Her work includes a VPH grism (a thick VPH grating sandwiched between two prisms constructed for the Subaru telescope in Hawaii), and a powerful online video matching application using optical correlation technology. She has also contributed greatly to nurturing many female scientists in Japan, and was granted the 2010 Prime Minister Award in recognition of this. She created two awards within the Japan Society of Applied Physics (JSAP: the Outstanding Female Researcher Award and the Contributions to Investing in People Award).

She became the Leader of the Committee for Promoting Equal Participation of Men and Women (part of JSAP) in 2001 and also headed the Japan Inter-Society Liaison Association Committee for Promoting Equal Participation of Men and Women in Science and Engineering (2002–2003). She was elected as the first female vice-president of the JSAP in 2006, and was a member of the Science Council of Japan (2005–2012). She has been a member of many government committees and advisory boards, including the Telecommunications Technology Council (Ministry of Internal Affairs and Communications). She is currently a member of Board of Directors of the Japan International Cooperation Center (JICE). She is a fellow of the SPIE (International Society for Optics and Photonics) and the JSAP.

(Co-author of Chapter 5) Takako Kodate (Doctorat en Sciences, spécialité informatique, Université de Nice-Sophia Antipolis, France) is currently a Lecturer in the Department of Information and Sciences, School of Arts and Sciences at Tokyo Woman's Christian University (TWCU).

Her main research area is graph theory and its applications to data transmission in wireless networks. She is conducting research to determine the best algorithm for data transmission under certain conditions given by a network modelled as a graph.

Her research interest also includes the roles of women-only universities, particularly for women in science. There are currently only five women's universities with a science faculty in Japan, and TWCU is one of them. She recently led a three-year project, 'Women's universities and the importance of their role in encouraging women in science', funded by the Institute for Women's Studies at TWCU (2010–2012). The results have been published in the project's final report (in Japanese). During this period, she has written two papers on a similar theme, together with the two authors of this book.

Acknowledgements

We are very grateful to anonymous reviewers and the Asian Studies team of Routledge for their comments and feedback as well as those who gave us invaluable information.

We would like to thank all those who have shared with us the goal of attaining a more equitable future for science and engineering, particularly the late Professor Masahide Kamiyama (Professor Emeritus, University of Tokyo), the late Professor Kōichi Kitazawa (former Executive Director, Japan Science and Technology Agency), Ms Atsuko Tōyama (former Minister of State for Education, Culture, Sports, Science and Technology), Ms Kumiko Bandō (Director General, Consumer Affairs Agency/former Deputy Director General, Ministry of Education, Culture, Sports, Science and Technology), Professor María J. Yzuel (Universitat Autònoma de Barcelona), and Professor Yoshikazu Tōyama (former Visiting Professor, Japan Women's University, JWU).

We are also grateful to those who provided advice and valuable comments during the preparation of this book, and for our previous research projects.

We gratefully acknowledge our mentors, colleagues and friends at JWU, the University of Tokyo, University College Dublin, London School of Economics and Political Science, Tokyo Woman's Christian University, Université de Nice-Sophia Antipolis, King's College London, the University of Electro-Communications, Japan, and Hokkaidō University.

Special thanks to all the alumnae, friends and supporters of the Kodate Laboratory, JWU (1972–2009), and all those who provided direct assistance for this book project: Ms Tomoko Nakayama, Hayanon and Dr Chiemi Fujikawa.

Last but not least, we would like to thank our parents, grandparents and family. This book is dedicated to all girls and women in STEM in Japan and the world, and future generations of scientists and engineers, both male and female.

Introduction

I.1 Gender equality and economy in Japan: where are we now?

In January 2014, Prime Minister Shinzō Abe gave a lecture at the Annual Meeting of the World Economic Forum in Davos, stating that 'Japan should be the place that gives women the opportunity to shine. Thirty per cent of people in leadership positions should be occupied by women by the year 2020'. It is significant that the Prime Minister chose an international forum to announce this as a priority issue.

The year 2015 marks the 20th anniversary of the Beijing Declaration and Platform for Action, signed by 189 governments at the 4th World Conference on Women. Since the first United Nations conference on Women was held in 1975, four conferences have been held, and four reviews carried out. They have served as catalysts for an international movement for women's empowerment and equality.

In response to these initiatives, national governments, business and civil society undertook various measures to promote the status of women and gender equality. In addition, each country is now compared against others based on indices such as the Human Development Index, Gender-related Development Index, and Gender Empowerment Measure (GEM).[1] In particular, the GEM underscores equality of opportunities in political and economic activities (e.g. the proportion of female legislators, senior officials, and managers, and women in professional and technical positions in each country).

The relationship between gender inequalities and economic growth has also attracted a wide range of research interest, from domains such as feminist studies, macroeconomics, labour market studies and social policy ([1] [2] [3] [4]).

Against this backdrop, gender equality in science, technology, engineering and mathematics (STEM) has been highlighted ([5] [6] [7] [8] [9] [10] [11])[2], and the goal of increasing the number of female scientists and engineers has become firmly ensconced in global policy agendas ([12] [13] [14]).

In Japan, gender equality has been promoted since the 1990s. It coincides with the period when the Japanese-style high economic growth model started to be questioned after the bubble economy burst and the Liberal Democratic Party (LDP) government collapsed. Throughout the 'lost decade' (1991–2000) ([15]), the concept of 'gender equality' was debated, particularly after the 4th World Conference on Women in 1995, and in 1999, the Basic Act for Gender Equal Society (BAGES) was enacted. Although the government and many stakeholders have been making efforts, Japan is still lagging behind other countries. For example, it ranked 104th among 142 countries in the most recent Global Gender Gap index (based on scores regarding health, educational attainment, political empowerment and economic participation). ([16])

A series of economic policies dubbed as Abenomics has been pursued under the current Abe government formed on 26 December 2012. Widely known as the 'three arrows', Abenomics consists of (i) bold monetary policy, (ii) flexible fiscal policy, and (iii) growth strategy. The third 'arrow' is targeted at encouraging women to be more actively engaged in society, particularly in the spheres of politics and the economy.

To indicate the government's strong will to lead the country, Ms Makiko Yamada was appointed as the very first female Executive Secretary to the Prime Minister in November 2013. This was the first female appointment to this position in its 128-year history ([17]). Likewise, in July 2014, four women were appointed to Directors-General in various government ministries, and the number of women appointed to executive or higher positions in the civil service (e.g. Director and Deputy Director General) doubled from 8 to 15 ([18]). Furthermore, the Prime Minister increased the number of female ministers in the Cabinet reshuffle in September 2014.

In parallel, the government announced a raft of policies to encourage women to be more actively involved in the labour market. Several targets were included in the government's revised Growth Strategy announced in June 2014. On top of the 30 per cent target for women in leadership positions, the government is also aiming to raise the women's labour participation rate (for women aged 25 to 44 years) from 68 to 73 per cent by 2020 ([19]).

In October 2014, the government submitted a bill dubbed 'Advancement of Women' (*Jyosei no katsuyaku soskushin hōan*)[3] to the parliament (the Japanese Diet, Kokkai). The bill requires national and local governments and companies with more than 300 employees to create an 'action plan' and publish their numerical targets on hiring and promoting women ([20]).

After nearly 30 years since the Equal Employment Opportunity Law (EEOL) was passed, Japan seems to be finally ready to welcome an environment in which an increasingly competitive global market operates.

Back in 1986, the EEOL was introduced to enable companies to fully utilise human resources regardless of gender. However, only a few firms (e.g. Japan IBM, Texas Instruments Japan) actively sought female talent. Into the 2000s, companies such as Tōshiba and Shiseidō began to implement equality and diversity policies. By 2003, the proportion of women in the head of unit role or equivalent positions in large corporations with 5000 or more employees was 5.2 per cent ([21]). Ten years later in 2013, the number increased to 10.2 per cent ([22]). The Ministry of Economy, Trade and Industry (METI) selects and awards companies with good practice in gender equality, while the Japan Business Federation ([23]) publicised information about some companies' policy and practice in promoting women. The Ministry of Health, Labour and Welfare (MHLW) is running an online portal to support working mothers. From January 2014, the Cabinet Office launched a website to present data from industry showing its support for working women ([24]).

The current government's policy on women and gender equality is therefore very much driven by its economic and active labour market policy. What about the situation surrounding Japanese women in scientific research? This is the issue that this book deals with.

I.2 Japanese women in science and engineering

In 1913, the first three female students were enrolled as full-time undergraduate students in a national university in the northern part of Japan. Interestingly, all of these students were in natural sciences (two in chemistry, and one in mathematics). Although one hundred years has passed since then, the proportion of female researchers is still one of the lowest (14.4 per cent as of 2013)[4] among OECD member states.

In 2010, a publishing company called Kōdansha Ltd. coined the word, '*Rikejo*' (abbreviation of 'rikei joshi', which literally means 'science girl'), and the term was popularised by several events and media reports. Rikejo is used to refer to girls (and women) who have an interest in 'scientific things', and it includes all women from secondary school to professional level (any occupation, no age limit), according to the firm's website. Many attempts have been made to break stereotypes and clichés about women in STEM, for example, by producing a comic and TV drama about a heroine who is scientifically gifted [27]). A newspaper in 2013 reported that there is now a minor surge in popularity among high school girls for selecting STEM programmes at the entrance exam for universities ([28]).

However, the recent positive trend has suffered a setback. In early 2014, it was announced that a female biologist at Japan's largest science research institute, *Rikagaku Kenkyūjo*, or RIKEN for short (the Institute of Physical and Chemical Research), made a breakthrough in finding a technique for creating cells that have the embryonic ability to turn into almost any cell type in the mammalian body ([29]). Initially, it looked as if the word *Rikejo* were going to be inseparable from this great scientific discovery, but it quickly turned into a scandal. Shortly afterwards, the researcher was accused of fabricating parts of scientific papers ([30]). Subsequently, the published articles were retracted in July 2014, and RIKEN came under severe scrutiny and criticism, particularly concerning the way in which they handled the case ([31]). After the original announcement in February, Professor Kaku Sechiyama (Gender Studies) challenged the perspectives taken by most of the Japanese media reports, which focused on the researcher's age and gender ([32]). He discovered that the same news reported in other Western countries did not even mention her age or gender. Japanese society's uneasiness with gender issues, and latent discrimination towards women continued to be discussed by the media in the following months ([33]).

While the gender gap in STEM varies greatly from country to country, no one would question the fact that Japanese women in these fields are still very few ([34] [35]). The above-mentioned recent incident and the gender-focused media reports underscored how Japanese women scientists are seen as rare, even in biology and life sciences.

In science, mathematics and computing, the ratio of female PhD graduates is 23 per cent in Japan, while their counterparts in the US and Spain account for 41 per cent and 48 per cent respectively. In engineering, manufacturing and construction, the ratio is almost half for each country (12 per cent in Japan, 24 per cent in the US and 34 per cent in Spain) ([13]) (Figure I.1).

By the mid-1990s, the underrepresentation of Japanese women scientists was recognised as an 'issue' by the Science Council of Japan (SCJ), which is a representative body of Japan's scientific community. When the 2nd Science and Technology Basic Plan was drawn up by the government in 2001, the problem of female underrepresentation in STEM was noted as a policy issue to be tackled.

To remedy the situation, the Japan Society of Applied Physics (ASAP), the Physical Society of Japan (PSJ) and the Chemical Society of Japan (CSJ) joined forces to form the Japan Inter-Society Liaison Association Committee for Promoting Equal Participation of Men and Women in Science and Engineering (EPMEWSE) in July 2002. EPMEWSE conducted a survey, commissioned by the Ministry of Education, Culture, Sports, Science and Technology (MEXT) and received almost 20,000 responses. The results were mentioned in the White Paper, and influenced subsequent policy decisions. Furthermore, in 2004, the government was hit by the news that Japan ranked lowest amongst the 30 OECD

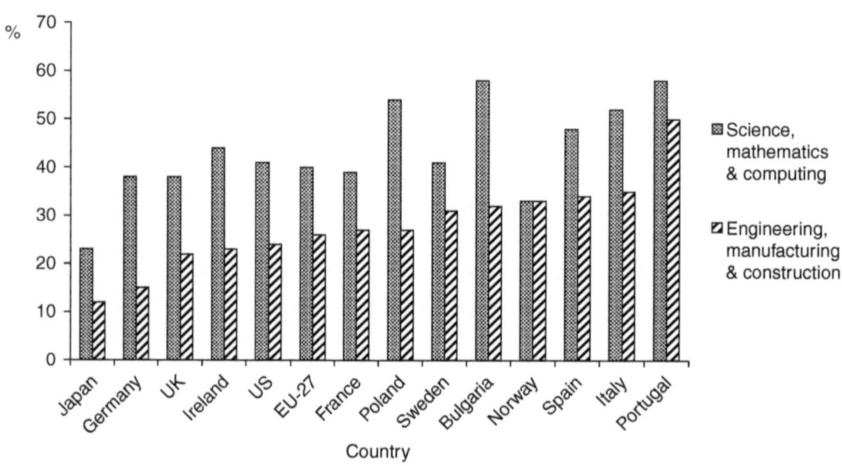

Figure I.1 Proportion of female PhD graduates in STEM, 2010 (European Commission, 2012) PhD graduates are defined as ISCED (International Standard Classification of Education) 6.

Source: [36]: 649.

member countries in terms of the percentage of women in the scientific workforce (11.1 per cent). Women account for 25–35 per cent in the other OECD countries.

In 2006, the government decided to set targets for the proportion of newly available permanent positions that would go to women in the 3rd Science and Technology Basic Plan in 2006. MEXT began to provide competitive grants, targeting universities and research institutes to promote the best model for fostering female researchers in STEM.

Galvanised by funding opportunities and clear direction from the government, universities and research institutes have been implementing a variety of measures. These organisations approach the issue of underrepresentation from the standpoint of systems support for career breaks for childbirth, and other life events such as child rearing and eldercare. These steady efforts have come to be seen as effective in changing the awareness of men and women, slowly but steadily, among staff and the next generation of researchers. Therefore, even before the current policy began to be targeted at companies and the civil service, the education and research sector in Japan has been engaged in promoting gender equality for nearly 10 years.

Japan is currently experiencing an unprecedented period of a declining and rapidly ageing population. Given that society has been based on strong patriarchy, a so-called 'male-breadwinner' model for such a long time, the government's present vision of a gender-equal society is radical, to say the least. If it is achieved while maintaining the country's strength in high-tech industries and basic science, it would mean a great transformation of Japanese society. Japan is at a crossroads, and women in STEM have been tasked with the great mission of turning the country's fate around.

I.3 Scope of this book: history and policy changes

Although there have been many policy changes since the 1990s, these changes in policy direction did not take place suddenly, but have had a long history. Women, families and organisations have tried to overcome barriers and hurdles, and striven for a change for many years in the past and even today. Although equal opportunity for men and women in education and employment has been legally recognised for decades, these formal opportunities were not necessarily taken up with enthusiasm until very recently. As a consequence, whilst female leaders began to emerge in various fields, they are still few and far between, particularly in STEM fields ([37]). Individual decisions, social and cultural norms, political and economic processes are all interwoven ([38]).

This book traces the history of Japanese women in STEM, covering key events and prominent figures in promoting women's access to science education at higher levels in Japan, while seeking to explain why the underrepresentation of female scientists went unchallenged throughout the duration of the post-war regime.

The book also explains how policy crystallised in the late 1990s/early 2000s. It sheds light on how gender equality in science and engineering policy was introduced through coordination among academia, scientific societies and government. The policy has yielded slow but steady progress in female representation to date. Although the number of female scientists in Japan is still relatively low, the book heralds a new era of female scientists by showcasing several programmes undertaken by universities and national research institutes to support multiple career paths for and the progression of female scientists.

Like many other countries, Japan is still in the middle of tackling the common problems known as 'glass ceiling' and 'leaky pipes' that are facing women in STEM (e.g. [39]). Yet there is a promising sign of younger female researchers balancing research and family life, while juggling with and overcoming barriers [40] [41]. This book seeks to capture these very significant recent developments as well as their history.

I.4 Japan's response to the 'why so few' question

International studies highlight the fact that the obstacles facing women in STEM are multifaceted and have to be carefully examined through the lens of their life-cycle stages ([42] [43] [44] [45]) (Figure I.2).

The slow progress in achieving gender equality in STEM has fuelled discussions about the underrepresentation of women in the field ('why so few') ([46] [45]), and the overall opinion is that in order to see desirable changes occur, a rather long-term process of implementing various reforms, including addressing biases in society, is required ([47]). In some countries where women are not encouraged to remain in the labour market, these changes may take longer than in other countries ([48]).

As Clark (2008) points out in a description of women's achievements and struggles in nineteenth-century Europe, women's access to a professional career was first restricted due to access to education being denied them for a long time. Once the equal right to education was granted to women, the next step was to open the door to a professional career: nursing, teaching, medicine, the legal profession, and public sector jobs. Countries vary greatly in the way in which women's access to various professions was regulated and gradually adjusted as women began trying to enter the professions. It also greatly mattered how society perceived women having a professional career.

Although most of the early female scientists in Japan chose to fulfil their duties as teachers while carrying out their research, the link between education and the prospect of a future career has always been important.

Just as more women are encouraged to be active in the labour market, some make a case for the introduction of quotas in order to engineer a rapid change in the numbers. Yet others argue against such a policy, because in order for women to have the desire to enter and pursue STEM in the long term, if not on a lifelong basis, 'there ought to be a mutual attraction between women and scientific fields' ([49]: 9). In recent years, a few research findings have started to emerge internationally showing the quality of science production being improved by addressing gender issues ([10], [50], [51], [52], [53], [54]). A number of case studies have

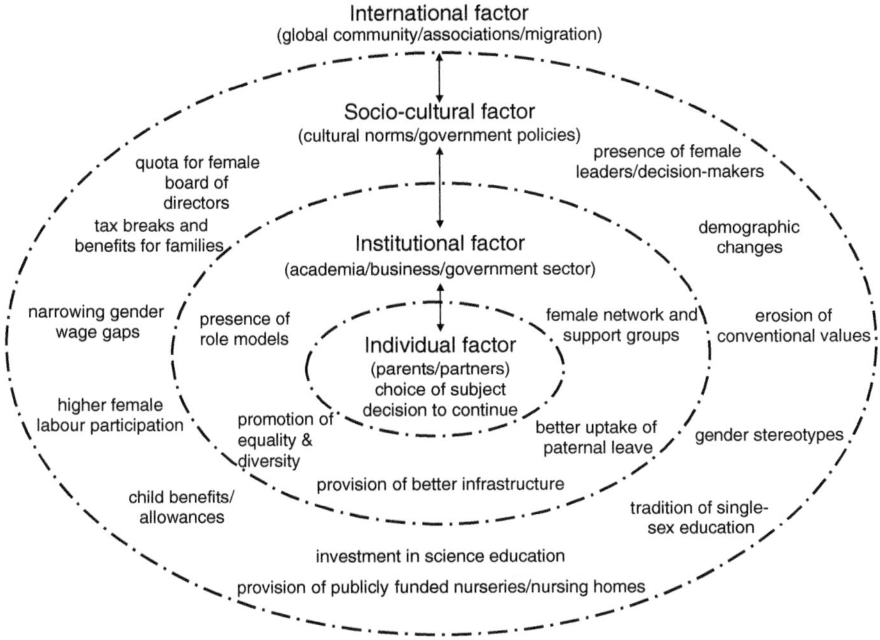

Figure I.2 STEM policies for women in socio-cultural, institutional and individual dimensions.

reported the positive effects of programmes successfully tackling gendered academic research cultures ([55], [56], [57], [58]).

Questions remain therefore about the extent to which government policies are committed to bringing about equality through the promotion of women in STEM and what their ultimate goals are, and about who plays a major role in implementing these policies ([59]). These are crucial questions for understanding not only the politics of gender equality, but also those of welfare states in transition, as the gender balance in these societies has major implications for education, for the economy and moreover, for the overall shape of each such country in the future (i.e. demographic changes) ([60] [61] [62] [63]).

This book seeks to address these questions by outlining the history and recent developments in Japan in terms of government and academic community initiatives for supporting female students and researchers in STEM. It explains how formal institutional arrangements for Japanese women in STEM were shaped at the end of the nineteenth century and in the early twentieth century, and were reformed at different 'critical junctures'[5] ([64] [65]).

This book is divided into three phases using those critical junctures (Figure I.3). The first phase deals with women's access to education and research in STEM. The second phase starts with the end of the Second World War (WWII), when women were allowed to enter the former Imperial Universities (*Teikoku Daigaku*), such as the University of Tokyo and Kyoto University. This phase ends in the mid-1980s,

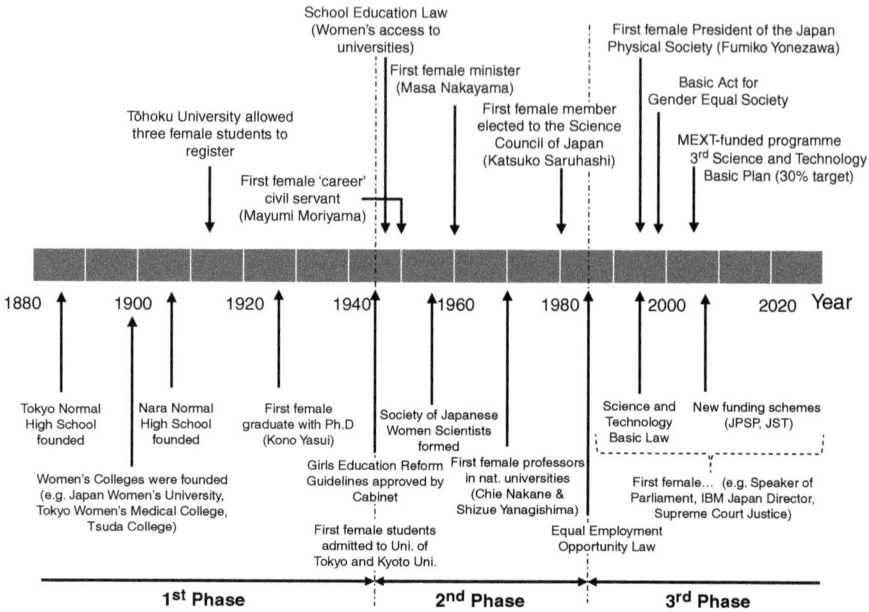

Figure I.3 Chronology and three phases in the history of Japanese women in science and engineering.

when equal opportunities at the workplace were rather half-heartedly forced upon companies. The third phase deals with the beginning of a gradual policy change in the Japanese political economy landscape and also gender equality in society.

Figure I.4 shows a conceptual map to explain the policy developments in terms of Japanese women in STEM.

This book aims to shed light on how policies for supporting and nurturing women in STEM have been guided by the concept of 'equal participation' in Japan. It places a particular focus on the interaction between government policies, institutional efforts and individuals. These three dimensions (national, organisational, and individual) have been identified as significant for explaining women's under representation at higher job levels ([66], [67]).

The book places an emphasis not only on the institutional constellations, but also on the role of ideas and public moods which affect government policy in a more informal manner. In Japan, the concept of 'gender equality' has not been widely prevalent and as strongly championed as in most Western feminist movements which have brought about changes in laws and social practice over many years. 'Gender equality' emphasises the importance of individual freedom from gender stereotypes. Although there have been several waves of women's movements[6] for social advancement in Japan, 'gender equality' has never been embraced by mainstream policymakers. The STEM policy was no exception.

In contrast, 'equal participation of men and women' (*Danjo Kyōdo Sankaku*) has been adopted widely since the early 1990s. Compared with 'gender equality',

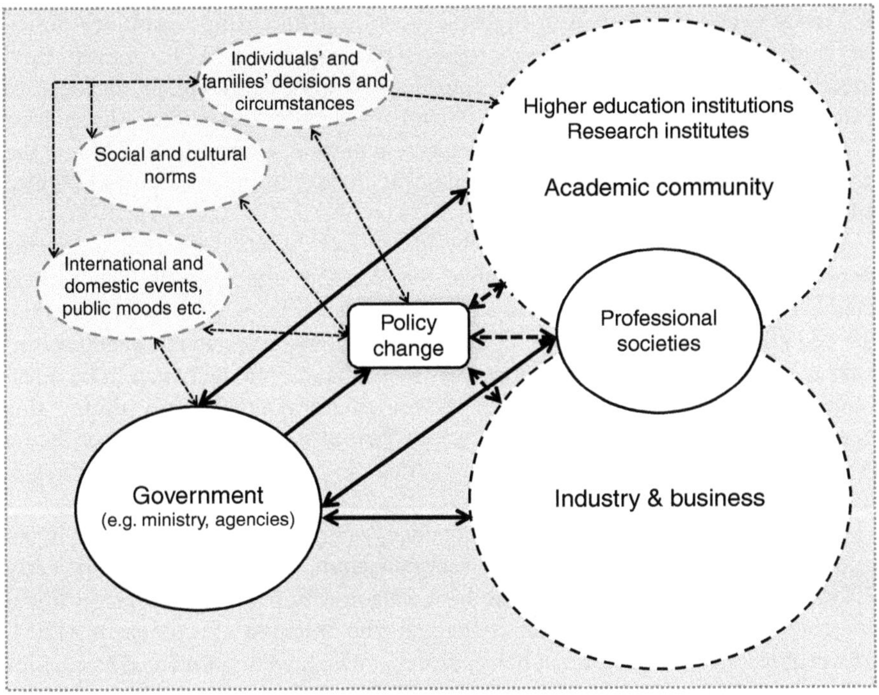

Figure I.4 Actors and factors that have affected gender equality policy reforms in STEM since the 1990s in Japan.

'equal participation' is more amenable to the long-held notion that women's roles at home as wives and mothers need to be valued equally with men's ([69]). The emphasis on the latter concept is based on a collective contribution to society as a group rather than in terms of individual achievements. This concept of 'equal participation' has played a major part in bringing about several major policy changes in the late 1990s and the early 2000s.

This book will serve as a case study of how the universal issue of underrepresentation of women in STEM has been tackled by national governments, which provided some unique programmes for supporting them on the ground in each university and national research institute. It consists of an introduction, five main chapters and a conclusion.

Chapter 1 describes the history of the establishment of the modern education system in Japan from the beginning of the Meiji era until WWII. The foundation of the Japanese school system will be outlined, with a particular focus on the development of women's access to higher and science education. It also discusses the concept of 'good wives, wise mothers', which has influenced the development of girls' education and schools.

Chapter 2 depicts the higher educational landscape in post-WWII Japan, highlighting the government's move towards a co-educational system, and the widening access of women to higher education. The chapter examines policy developments in terms of women's rights, leading up to the EEOL in 1986. Both the EEOL and co-education were introduced rather passively in response to external pressure, and not warmly welcomed by many stakeholders in society. The significant problem of the underrepresentation of women in STEM and the activities of academic societies in scientific fields during that period is also addressed.

Chapter 3 looks at the impacts of the EEOL and the developments during the period between 1986 and 1999, when the BAGES was enacted. The Liberal Democratic Party (LDP) governed Japan for 38 years from 1955 and 1993 when the so-called 1955 regime came to an end. Simultaneously, as the bubble economy burst, many attributes of the post-WWII Japanese model began to be questioned. The educational attainment of women grew significantly during this period, and more women enrolled in a four-year university course, rather than a two-year college course. The meaning of education and its link to a career began to change for women.

Chapter 4 describes the two important changes prior to the policy change in the mid-2000s. One is the government reorganisation, which took place in early 2000, and another is the formation of a liaison group of various professional societies in order to bring about changes to the situation of women in STEM. The chapter then outlines a number of policies which were introduced to ameliorate the underrepresentation of women in STEM.

Chapter 5 focuses on the evaluation of government policies in the last 8 years, based on the questionnaire results of professional societies and those of higher education institutions. The voices of women in STEM (alumnae from universities, and those active in research and industry) are also highlighted, in order to contrast them with the focus of macro-level policies on the 'why so few' (quantity) question. The final chapter concludes by looking at current and future challenges and opportunities facing Japanese women in STEM.

Although this book is concerned with gender equality in science education and research, its scope is limited to STEM subjects, and does not include medicine, nursing and pharmacy. Moreover, the primary focus is on women scientists and engineers in the education and research sector, rather than those in industry. While there are also many works put forward by scholars in gender studies regarding this subject, the authors have sought to put the issue in its historical context and examine it from the perspective of public policy. Therefore, the book will be of interest to students and researchers from a wide range of disciplines, including: Asian studies, comparative social policy, history of science and technology, sociology, gender studies, political science, and public policy.

Notes

1 The Geneva-based international non-profit organisation, the World Economic Forum, has published the *Global Gender Gap Report* annually since 2005, which covers and ranks 130 major and emerging economies. Thirteen out of the fourteen variables used for assessing inequality between men and women are based on publicly available

data from the International Labour Organisation, the United Nations Development Programme and the World Health Organisation.

2 In the United States, the first report on women and minorities in science and engineering was published by the National Science Foundation (NSF) in 1982.

3 The official title is 'Bill on comprehensive and intensive promotion of the development of social environment where women can flourish' (*Jyosei ga katsuyaku dekiru shakai kankyo no seibi no sōgōteki katsu shūchūteki na sokushin ni kansuru hōritsuan*). However, on 21st November 2014, the House of Representatives was dissolved, and a snap election was held on 14th December 2014. As a result this bill was not passed.

4 cf. 17.6 per cent in South Korea (2011), 25.6 per cent in France (2010) and 38.3 per cent in the UK (2010) ([25] [26]).

5 Critical junctures are defined as 'a period of significant change, which typically occurs in distinct ways in different countries (or in other units of analysis) and which is hypothesized to produce distinct legacies' ([64]: 29).

6 One of the most well-known women's liberation movements in the 1970s is represented by *'Gurūpu Tatakau Onnatachi' (Fighting Women Group)*, led by Mitsu Tanaka ([68]).

References

[1] D. Robbins. Gender, Human Capital and Growth: Evidence from Six Latin American Countries. Technical Report 151, OECD Development Centre Working Paper, 1999. Paris: OECD.

[2] S. Klasen. Low schooling for girls, slower growth for all? Cross-country evidence on the effect of gender inequality in education on economic development. *The World Bank Economic Review*, 16(3): 345–373, 2002.

[3] S. Seguino and C. Grown. Gender equity and globalization. *Journal of International Development*, 18(8): 1081–1104, 2006.

[4] J. Jütting and C. Morrisson. L'égalité entre les sexes est un outil du développement (Gender Equality is a Development Tool) [In French]. *Le Monde*, 13 Jan 2006.

[5] European Commission. Addressing the gender balance in research policy. [online]. Available from: http://cordis.europa.eu/etan/src/topic-4.htm. Accessed 27 November 2014.

[6] H. Etzkowitz, C. Kemelgor, B. Uzzi, and M. Neushatz. *Athena Unbound: The Advancement of Women in Science and Technology*. Cambridge: Cambridge University Press, 2000.

[7] S. Harding. *Whose Science? Whose Knowledge? Thinking from Women's Lives*. Ithaca, NY: Cornell University Press, 1991.

[8] H. Longino. *Feminist Epistemologies*, chapter 'Subjects, power and knowledge: description and prescription in feminist philosophies of science'. New York: Routledge, 1993. eds. Linda Alcoff and Elizabeth Potter.

[9] OECD. *Women in Scientific Careers: Unleashing the Potential*. Paris: OECD, 2006.

[10] L. Schiebinger. *Has Feminism Changed Science?* Cambridge, MA: Harvard University Press, 1999.

[11] A. Tsugawa and Y. Kanomi. *Hiraku: Nihon no josei kagakusha no kiseki (Opening: Tracing the paths of Japanese female scientists)*. [In Japanese] Domesu Publishing, 1996.

[12] L. Schiebinger. Gender, Science and Technology. Background paper submitted to Expert Group Meeting at United Nations Division for the Advancement of Women

(DAW, part of UN Women) and United Nations Educational, Scientific and Cultural Organization. [online]. Available from: http://www.un.org/womenwatch/daw/egm/gst_2010/Schiebinger-BP.1-EGM-ST.pdf, 2010. Accessed 9 November 2014.

[13] European Commission. She Figures 2012: Gender in Research and Innovation. Technical report, Brussels: Director-General for Research, 2012. [online]. Available from: http://ec.europa.eu/research/science-society/document_library/pdf_06/she-figures-2012_en.pdf. Accessed 9 November 2014.

[14] UNESCO (2013). *IGU/UNESCO Workshop on Women in Engineering*. [online] Available from: http://www.unesco.org/new/en/natural-sciences/science-technology/engineering/women-in-engineering/workshop-on-women-in-engineering/. Accessed 9 November 2014.

[15] F. Hayashi and E.C. Prescott (2003). The 1990s in Japan: A lost decade. [online]. Available from: http://fhayashi.fc2web.com/Prescott1/Postscript_2003/hayashi-prescott.pdf. Accessed 10 April 2014.

[16] World Economic Forum. Technical report, Geneva: World Economic Forum, 2014. [online]. Available from: http://www3.weforum.org/docs/GGGR14/GGGR_CompleteReport_2014.pdf Accessed 9 November 2014.

[17] *Nikkei Shinbun*. Kasumigaseki, hipparu nowa watashi. Gaikō/kyōiku de jyosei kanryō funtōki. (It is me who is going to lead Kasumigaseki (Government Departments). How rising women civil servants in diplomacy and education are juggling work and family life) [In Japanese]. [online]. 21 December 2013. Available from: http://www.nikkei.com/article/DGXNASDF1300F_T11C13A2TY5000/, 2013. Accessed 9 November 2014.

[18] *Asahi Shinbum. Shōchō kanbu, jyosei baizō, naikaku jinji kyoku, seiken no ikō han-ei.* (female staff at executive levels doubled in ministries, as the Cabinet Bureau of Personnel Affair's decision reflects government's intention) [In Japanese]. [online] 5 July 2014. Available from: http://digital.asahi.com/articles/DA3S11225865.html, 2014. Accessed 9 November 2014.

[19] Nippon.com. Jyosei no katsuyaku sokushin. 'Womanomics' de nihon ga kawaru? (Advancement of women: Will 'womanomics' change Japan?) [In Japanese]. [online]. Available from: http://www.nippon.com/ja/features/h00066/, (n.d.). Accessed 15 November 2014.

[20] House of Representatives. Jyosei ga katsuyaku dekiru shakai kankyō no seibi no sōgōteki katsu shūchūteki na sokushin ni kansuru hōritsuan (Bill on comprehensive and intensive promotion of the development of social environment where women can flourish). [online]. Available from: http://www.shugiin.go.jp/internet/itdb_gian.nsf/html/gian/honbun/houan/g18601038.htm, 2014. Accessed 15 November 2014.

[21] Ministry of Health, Labour and Welfare. Danjyo koyō kikai kintō kankei shiryō. (Data regarding Equal Employment Opportunity) [In Japanese]. [online] Available from: http://www.mhlw.go.jp/shingi/2004/09/dl/s0914-6e2.pdf, 2004. Accessed 11 November 2014.

[22] Ministry of Health, Labour and Welfare. Course betsu koyō kanri seido no jisshi/shidō jyōkyō (Investigation into the dual-track career system) [In Japanese]. [online] Available from: http://www.mhlw.go.jp/general/seido/koyou/danjokintou/dl/cource_joukyou.pdf, 2013. Accessed 15 November 2014.

[23] Keidanren Japan Business Federation (2014). Jyosei no yakuin/kanrishoku tōyō ni kansuru jisyu kōdō keikaku (Voluntary action plan on appointment of women executives and managers). [In Japanese]. [online]. Available from: http://www.keidanren.or.jp/policy/woman/actionplan.html. Accessed 15 November 2014.

[24] Cabinet Office. Jyosei no katsuyaku 'mieru-ka' saito (Website for 'visualising' advancement of women) [In Japanese]. [online] Available from: http://www.gender.go.jp/policy/mieruka/company/mierukasite.html, 2004. Accessed 15 November 2014.

[25] Ministry of Internal Affairs and Communications Statistics Bureau and Communications (2013). *Kagaku gijyutsu kenkyū chōsa hōkoku* (Study report on science and technology research) [In Japanese]. [online]. Available from: http://www.stat.go.jp/data/kagaku/kekka/index.htm. Accessed 15 November 2014.

[26] OECD. Main Science and Technology Indicators. [online]. Available from: http://www.oecd.org/sti/msti.htm, 2014. Accessed 12 November 2014.

[27] *New York Times*. Japan's 'Science Women' seek an identity. [online]. Available from: http://www.nytimes.com/2013/06/17/world/asia/Japans-Science-Women-Seek-an-Identity.html?pagewanted=all&_r=0, 16 June 2013. Accessed 28 November 2014.

[28] *Nikkei Shinbun*. 'Rikejo' wa nihon no monodukuri wo kaeruka? Kanojo tachi no honne to kigyō no omowaku (Can 'Rikejo' change manufacturing in Japan? Their honest opinions and companies' intentions) [In Japanese]. [online]. 6 November 2013. Available from: http://business.nikkeibp.co.jp/article/interview/20131030/255343/?rt=nocnt, 2013. Accessed 9 November 2014.

[29] D. Cyranoski. Acid bath offers easy path to step cells. *Nature*, 29 January 2014. [online] Available from: http://www.nature.com/news/acid-bath-offers-easy-path-to-stem-cells-1.14600#/b2 Accessed 9 November 2014.

[30] K. Takenaka (2014). Japan finds fraudulent steps in 'breakthrough' stem cell paper. *Reuters*. [online]. 1 April 2014. Available from: http://www.reuters.com/article/2014/04/01/us-stemcells-idUSBREA300PW20140401. Accessed 9 November 2014.

[31] M. Maruko (2014). After STAP scandal, Riken plots its future. *Japan Times*, 1 September 2014. Available from: http://www.japantimes.co.jp/news/2014/09/01/reference/stap-scandal-riken-plots-future/#.VF-sQfnz2So. Accessed 9 November 2014.

[32] K. Sechiyama (2014). STAP saibō 'jyosei/sanjyū sai' hōdō wa Nikkan dake? Naze nenrei to seibetsu ga korehodo jyūshi sarete shimau noka (Media reports on STAP cell breakthroughs: 'woman/30 years old' only found in Japanese and Korean media. Why does age and gender attract so much attention?) [In Japanese]. [online] *Tōyō Keizai Online*, 13 February 2014. Available from: http://toyokeizai.net/articles/-/30422?page=3. Accessed 9 November 2014.

[33] M. Iwamoto and M. Takahashi. Onna to otoko, STAP sōdō kara kangaeta. kakureta ishiki ga hataraku toki. (Women and Men: When Hidden Consciousness Shows. What STAP Scandal Tells Us) [In Japanese]. *Asahi Shinbum* [online]. 9 July 2014. Available from: http://digital.asahi.com/articles/ASG6R3PHDG6RUTFK003.html?iref=com_rnavi_arank_nr03, 2014. Accessed 9 November 2014.

[34] Japan Times. Japan's scientists: just 14% female. Lowest rate among developed nations is a loss for the economy. [online] Available from: http://www.japantimes.co.jp/news/2014/04/15/national/japans-scientists-just-14-female/#.VHi5ENKDlNc, 2014. Accessed 28 November 2014.

[35] Times Higher Education. *Ambitious female scientists leave Japan to escape 'male domination'*. [online] Available from: https://www.insidehighered.com/news/2014/10/09/japanese-university-president-laments-exodus-women-science, 9 October 2014. Accessed 28 November 2014.

[36] N. Kodate, K. Kodate, and T. Kodate. Paving the way and passing the torch: mentors' motivation and experience of supporting women in optical engineering. *European Journal of Engineering Education*, 39(6): 648–665, 2014.

[37] L. Whipp. Japan: culture change required. Women at the top. *Financial Times*, December 7, 2011.

[38] F. Bray. *Technology and Gender: Fabrics of Power in Late Imperial China*. Berkeley: University of California Press, 1997.

[39] D. Bilimoria and X. Liang. *Women, Science, and Technology: A Reader in Feminist Science Studies*, chapter 'State of knowledge about the workforce and participation, equity, and inclusion of women in academic science and engineering', pages 21–50. New York: Routledge, third edition edition, 2014. eds. M. Wyer, M. Berbercheck, D. Cookmeyer, H.Ö. Öztürk, and M. Wayne.

[40] A.S. Aronsson. *Career Women in Comtemporary Japan: Pursuing Identities, Fashioning Lives*. New York: Routledge, 2014.

[41] *Asahi Shinbum Weekly AERA*. Rikejo no jidai (Era of science women) [In Japanese], 24 February 2014.

[42] J.C. Blickenstaff. Women and science careers: Leaky pipeline or gender filter? *Gender and Education*, 17: 369–386, 2005.

[43] C. Castaño and J. Webster. Understanding women's presence in ICT: the life course perspective. *International Journal of Gender, Science and Technology*, 3(2): 365–386, 2011.

[44] European Commission. *Women in Science and Technology: Creating Sustainable Careers*. Brussels: Director-General for Research, 2009.

[45] C. Herman and J. Webster. Taking a life-cycle approach: redefining women returners to science, engineering and technology. *International Journal of Gender Science and Technology*, 2(2): 179–205, 2010.

[46] C. Wennerås and A. Wold. Nepotism and sexism in peer-review. *Nature* 387, 22. May 1997. 341–343.

[47] C.A. Moss-Racusin, J.F. Dovidio, V.L. Brescoll, M.J. Graham, and J. Handelsman. Science faculty's subtle gender biases favour male students. *Proceedings of the National Academy of Sciences of the United States of America*, 109(41): 16474–16479, 2008.

[48] O. Sjöberg. Ambivalent attitudes, contradictory institutions: Ambivalence in gender-role attitudes in comparative perspective. *International Journal of Comparative Sociology*, 51(1–2): 33–57, 2010.

[49] M. Bouville. Should there be more women in science and engineering? [online] Available from: http://arxiv.org/pdf/physics/0611089.pdf, 2006. Accessed 15 November 2014.

[50] S. Ihsen, S. Gebauer, and V. Hantschel. *Gender and Education: Towards New Strategies of Leadership and Power*, chapter 'The relevance of gender issues in science and engineering education', pages 63–67. Berlin: Druck und Verlag, 2011, eds. Hildegard Macha, Claudia Fahrenwald, and Quirin J. Bauer.

[51] S. Le-May Sheffield. *Women and Science: Social Impact and Interaction*. New Brunswick, NJ: Rutgers University Press, 2006.

[52] L. Osborne, K. Miller, and R. Farabee-Siers. *Pedagogical methods for improving women's participation and success in engineering education*. Washington, DC: Institute for Women's Policy Research, 2008.

[53] Svein Sjøberg and Camilla Schreiner. *The ROSE project. Overview and key findings*. Oslo: University of Oslo, 2010.

[54] M. Uriate, H.A. Ewing, V.T. Eviner, and K.C. Weathers. Constructing a broader and more inclusive value system in science. *BioScience*, 57(1): 71–78, 2007.

[55] J.C. Sible, D.E. Wilhelm, and M. Lederman. Teaching cell and molecular biology for gender equity. *CBE Life Sciences Education*, (5): 227–238, 2006.

[56] M.F. Fox. *Higher Education: Handbook of Theory and Research*, chapter 'Institutional transformation and the advancement of women's faculty. The case of academic science and engineering', pages 73–103. Dordrecht, Germany: Springer Science + Business Media B.V., 2008. edited by John C. Smart.

[57] M.F. Fox, G. Sonnert, and I. Nikiforova. Successful programs for undergraduate women in science and engineering: adapting versus adopting the institutional environment. *Research in Higher Education*, (50): 333–353, 2009.

[58] L.A. Rhoton. Distancing as a gendered barrier: Understanding women scientists' gender practices. *Gender and Society*, 25(6): 696–716, 2011.

[59] N. Kodate, K. Kodate, and T. Kodate. Mission completed? Changing visibility of women's colleges in England and Japan and their roles in promoting gender equality in science. *Minerva*, 48(3): 309–330, 2010.

[60] J. O'Conner, A. Orloff, and S. Shaver. *States, Markets, Families: Gender, Liberalism and Social Policy in Australia, Canada, Great Britain, and the United States*. New York: Cambridge University Press, 1999.

[61] M. Estévez-Abe, I. Torben, and D. Soskice. *Varieties of Capitalism: The Institutional Foundations of Comparative Advantage, eds. Peter Hall, and David Soskice*, chapter 'Social protection and the formation of skills: A reinterpretation of the welfare state', pages 145–183. New York: Oxford University Press, 2001.

[62] F. Rosenbluth T. Iversen and D. Soskice. Divorce and the gender division of labor in comparative perspective. *Social Politics*, 12(2): 216–242, 2005.

[63] L. McCall and A. Orloff. Introduction to special issue of social politics: Gender, class and capitalism. *Social Politics*, 12(2): 159–169, 2005.

[64] R.B. Collier and D. Collier. *Shaping the Political Arena: Critical Junctures, the Labour Movement and Regime Dynamics in Latin America*. Princeton, NJ: Princeton University Press, 1991.

[65] P. Pierson. *Politics in Time, History, Institutions and Social Analysis*. Princeton, NJ: Princeton University Press, 2004.

[66] B.R. Ragins and E. Sundstrom. Gender and power in organizations: A longitudinal perspective. *Psychological Bulletin*, 105(1): 51–88, 1989.

[67] E.A. Fagenson. At the heart of women in management research: theoretical and methodological approaches and their biases. *Journal of Business Ethics*, 9(4–5): 278–274, 1990.

[68] A. Mizoguchi, Y. Saeki, and S. Miki. *Shiryō nihon ūman ribu shi (Documentation of History of Women's Liberation in Japan)*. [In Japanese] Kyoto: Shokadō Shoten Publishing, 2008.

[69] S. Buckley. *Broken silences: Voices of Japanese feminism*, chapter 'Ueno Chizuko', pages 272–301. Berkley: University of California Press, 1997.

1 History of women's education in pre-WWII Japan

1.1 Introduction

This chapter covers the history of the establishment of the modern education system in Japan from the beginning of the Meiji era until the Second World War. It sheds light on the foundation and evolution of the Japanese school system, with a particular focus on the development of women's access to higher and science education. It also discusses the concept of 'good wives, wise mothers' (*ryōsai kenbo*), which influenced the development of girls' education and schools. Part of the government's policy included, the establishment of Higher Normal Schools as the highest education institutions for women, and the further development of women-only colleges in the private sector. Emerging from a non-existent or narrow path, some women managed to become scientists and even obtained doctoral degrees by the late 1920s. The last section looks at some women's movements of that time and their achievements leading up to WWII.

Emphasis is placed on the three factors that seem absolutely necessary for the advancement of women in STEM in Japan. The first factor is a high level of sensitivity towards Western values that was equated with the modernisation of the country by the government. The second factor is the extent to which social and cultural norms were questioned by social campaigns at the time. The third factor is the level of support and consensus among the academic community regarding the issue of gender inequality within STEM fields.

1.2 First dawn: the Meiji Restoration and Western influence

Like many countries in the Western world, women in Japan had no access to higher education until the beginning of the twentieth century. As the Meiji government gained power in 1868, many reformist ideas were brought into the country from the West ([1]) and gradually implemented.

In 1871, the Ministry of Education was established, which in turn set up a school system based closely on the American model. Several bills were introduced with the intention of providing education for every citizen. The first School System Order (*Gakusei*, 1873) led to the creation of a 'university' (*daigaku*) for the first time as the higher institution for men's education. The five-year primary school was made compulsory for all children in 1872. These elementary schools were designed to create loyal subjects of the Emperor. However, tuition fees for primary schools prevented many families from sending their children to school until 1900, when fees were abolished. The gap between boys and girls in education was also notable: 72 per cent of boys as opposed to only 37 per cent of girls in 1892. The gap decreased with the development of the compulsory school system. By 1907, when the six-year primary school became compulsory, attendance figures for boys and girls for primary education were almost equal at over 95 per cent each ([2]).

Middle Schools were created as preparatory schools for students destined to enter one of the Imperial Universities, which were intended to produce leaders able to direct the modernisation of Japan. Since women were not allowed to enter the Imperial Universities, middle schools only targeted boys.

In 1879, another law, the Education Order, was enacted as Dajōkan Fukoku (Edict or Proclamation of the Grand Council of State), and replaced the original 1873 School System Order. This Order allocated greater discretion to local authorities, thus abolishing the 1873 School System and replacing it with 'school districts' which were controlled at the local level of towns and villages. Local authorities became the administrators of educational 'districts'. With this Order, 'schools' (*gakkō*) were classified into six categories: primary school (*shōgakkō*), middle school (*chūgakkō*), university (*daigaku*), normal school (*shihan gakkō*), vocational school (*senmon gakkō*), and 'others'. This Order consisted of 47 articles and provided a general framework for other more specific laws[1] which were subsequently enacted for each category. The following figure shows the school system in the pre-WWII period (as of 1935).

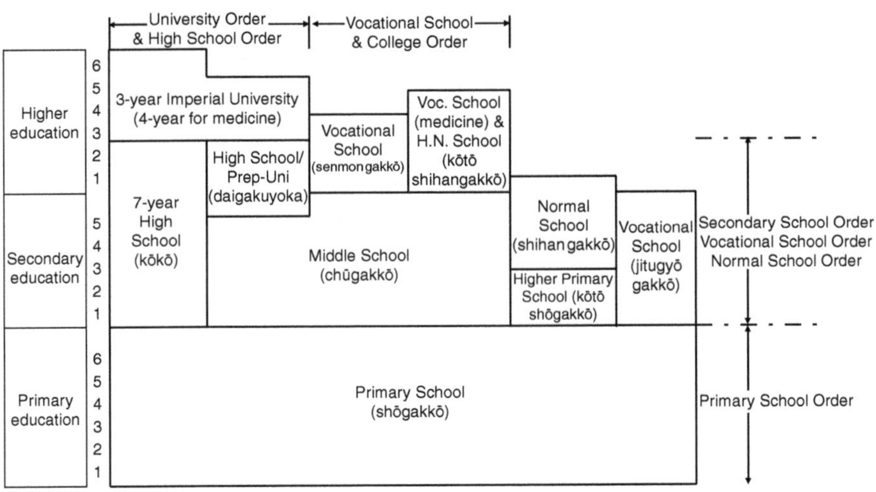

Figure 1.1 School system in pre-WWII Japan.

Source: [3] Adapted by the authors.

During this period, middle schools went through a series of reforms under the auspices of the then Education Minister, Arinori Mori. The 1886 Middle School Order created two types of middle schools: *Kōtō-chūgakkō* (only five schools across the country, administered by the Ministry of Education) and *Jinjō-chūgakkō* (administered by the prefectures). The Women's Normal School (*Joshi-shihangakkō*) had been open since 1875, and not only prepared women for teaching positions but also became the national centre for girls' education ([4]). In 1882, its preparatory school was transformed into Higher School for Girls (*Kōtō-jogakkō*). However, the 1886 Middle School Order made no mention of girls' education. Some amendments were made in 1891, and Higher School for Girls was classified as being of the type *Jinjō-chūgakkō* in the amended Order.

In 1899, a separate Order for Higher Schools for Girls (Kōtō-jogakkō) was enacted. It stipulated that the girls eligible to enrol in this school were those who had completed the higher grades of primary school (and were over 12 years of age). Each prefecture had the responsibility for planning and providing this type of school for their residents. It was left up to each individual prefecture to decide on the total number of school years for this type of school, which ranged from three to five. With the 1910 amendments, these schools were permitted to establish home economics-based specialised courses, since the primary aim of education for women was to produce 'good wives and wise mothers' (*Ryōsai kenbo*). The curriculum emphasised conservative, traditional Japanese values such as 'men as primary family breadwinners' and 'women as full-time housewives' and 'good wives'. From this gender-distinct curriculum, we can surmise that the system was clearly geared towards boys and thus considered unsuitable for girls. In addition, schools had to adapt themselves to the needs of family lives, by creating a class especially for girls in order for them to acquire practical skills such as sewing and cooking ([2]).

Although Mori's devotion to education for boys and girls has been noted and influenced national policies, he had a clear idea that the whole purpose of female education should lie in fostering 'good wives, wise mothers'. The term 'good wives, wise mothers' was coined by a pioneer in women's education, Masanao Nakamura in 1875, and began to represent the ideal for womanhood in the late 1880s and early 1900s ([19]). The original idea had much deeper historical roots. It derived from an old Confucian idea and the well-known textbook, 'Greater Learning for Women', written in the Tokugawa era (1600–1867). Although its direct impact on the new education system under the Meiji government is unclear, Sekiguchi ([20]: 14), recognises based on her analysis of historical archives, that not only Western thought but also some Confucian ideas had an influence on the Peers School for Girls (*Kazoku Jogakkō*) that was inaugurated in 1885. Through the new education system, the notion 'men work outside the home, while women stay at home' became a dominant social norm in Japan.

Mori developed policy on how to educate girls and distributed the 'Guidelines for educating [female] pupils' to girls' schools and girls' high schools nationwide in 1885. The notion of 'good wives, wise mothers' had a lasting/long-term effect on the national education policy for girls and women.

1.3 Looking for the best model for female education

In 1871, the Iwakura Embassy (Iwakura Shisetsudan) to the West was dispatched, initially leaving Yokohama for San Francisco. It was launched by the government with the goal of modernising Japan based on a Western social and economic model. 'Modernisation' was considered to be essential for a renegotiation of the unequal treaties that existed with the United States and Europe at that time. In the 1850s, the Japanese government resumed international trading with other Western countries than the Netherlands, and signed a treaty that granted the Western powers (the United States, the United Kingdom, France, Russia and the Netherlands) certain privileges. It is collectively known as Ansei Five-Power Treaties. The unequal aspects were found primarily in two aspects: a system of extra-territoriality (where foreign residents were subjugated to the laws of their own courts instead of the Japanese legal system), and fixed low import–export duties (which meant that the Japanese government had no control of foreign trade and protection of national industries).

The Iwakura Embassy's goal of modernisation was to end such unequal treatment. It visited the US, twelve European Countries including the UK, France, Belgium, and Prussia and a couple of Asian countries such as Singapore and Hong Kong.

The importance of girls' education was recognised as Iwakura's Embassy consisted of fifty-eight members, of whom five were women. However, a clear distinction was made between men's and women's education. The purpose of sending the male students was to equip them as quickly as possible with all the skills and knowledge necessary for modernising the country upon their return to Japan, while the five female students sent were aged seven to fourteen. The area of female education was not a priority for the government, as demonstrated by the fact that these ladies were relatively younger than their male counterparts. Two of the five female members returned to Japan for health-related reasons one year later, and two others (Shigeko Uryū (née Nagai) and Sutematsu Ōyama (née Yamakawa)) married after their ten-year study-abroad period in 1881. The only woman who remained unmarried and dedicated her life to women's education was Umeko Tsuda. She was seven years old, the youngest member of the party at the start of the expedition in 1871, and came back to Japan in 1882 at the age of eighteen. Since her Japanese language ability had significantly decreased, she faced difficulty readjusting to Japanese society. She decided to return to the United States, with a clear vision of devoting herself to women's education. She attended Bryn Mawr College in Philadelphia from 1889–1892 and studied biology and education. She also went to the United Kingdom to study at St Hilda's College, Oxford. After her return to Japan, she founded the Women's Institute for English Studies (*Joshi Eigaku-juku*) in Tokyo, which later became Tsuda College ([5]: 181–183). The other two (Uryū and Ōyama) were equally committed to women's education. Both graduated from Vassar College, and after their return to Japan, while Uryū taught music at Tokyo Women's Higher Normal School (see the following section) and the Tokyo Music College (now Tokyo University of the Arts), Ōyama contributed to the establishment of Peers' School for Girls (which later became Gakushūin Women's College, in 1906). Ōyama also supported Tsuda in the foundation of Tsuda College.

1.4 Female education in the Meiji era and the first universities in Japan

Although progress was made in women's education, the overall social system was reformed on the lines of 'Enriching the country, strengthening the military' (*Fukoku Kyōhei*) and 'Promoting new industries' (*Shokusan Kōgyō*). A number of universities were founded with the clear mission of importing and translating Western ideas into Japan in order to catch up with the West. In 1877, the University of Tokyo was established with four faculties (Law, Humanities, Science and Medicine) and became the Imperial University in 1886 by the enactment of the *Order of the Imperial University*. At this stage, it was the only Imperial University in Japan, absorbing the former Imperial College of Engineering.

Prior to 1886, the university (*daigaku*) was not a major player in higher education in Japan. Instead, many publicly established colleges such as Sapporo Agricultural College and the Imperial College of Engineering existed to provide education to students. The Order of the Imperial University in 1886 was a catalyst that changed the landscape of higher education in Japan. It set the scene for the establishment of six other Imperial Universities[2] across the country as Japan's top-level education organisations, while simultaneously other types of higher education institutions were built. The hierarchical structure within the higher education system in Japan has its roots in these developments.

The establishment of these Imperial Universities was an integral part of the grand scheme to foster the elites for modern Japan ([6]: 131–143). With the increasing industrialisation of Japan, the demand for educated and skilled workers also increased. Higher education and vocational training were both deemed necessary, and so the then Minister of Education Kowashi Inoue decided to found state-funded vocational schools, while also promoting women's education through a separate single-sex schooling system.

Amid these projects, education for women developed through the establishment of the Tokyo Women's Normal School in 1875. For a brief period during 1885, it was merged into the Tokyo Higher Normal School as its girls' branch. However, it was upgraded to college status when it became Tokyo Women's Higher Normal School (TWHNS) in 1890 (and was subsequently transformed into its current form as Ochanomizu University in 1949). In 1908, another Women's Higher Normal School was established in Nara. These two Women's Higher Normal Schools played a major role as the highest level of educational institution for women until the end of WWII, as access to further education for women was considerably restricted otherwise. Women were not allowed to study at any of the Imperial Universities until 1913.

Another type of higher education institution for women was initiated by several individual educationalists and founded as private enterprises called Colleges. They were often set up under Christian guiding principles, and maintained a Christian ethos. As a private educational institution, American Presbyterian missionaries opened a school of English in 1870, which was the forerunner to Ferris University. Umeko Tsuda founded the Women's Institute for English Studies in 1900, as previously mentioned. Tsuda had been baptised as a Christian during her stay in the USA. Although Tsuda College was not established as a missionary school, there is a chapel on the campus.

In the same year, Yayoi Yoshioka (who was a female doctor) decided to found a women-only medical school (later known as Tokyo Women's Medical University) after hearing the news that a woman had been denied admission to Saiseigaku-sha (precursor to Nippon Medical University).[3]

In 1901, the Japan Women's University (JWU) was established by Jinzō Naruse, who was a Christian church minister. It was originally recognised by the government as a 'college', not a University. With the idea initiated by a graduate of Tokyo Women's Higher Normal School, Tetsu Yasui, and A.K. Reischauer (father of Edwin O. Reischauer), Tokyo Woman's Christian University (TWCU) was founded by Inazō Nitobe as a 'college' in 1918. Nitobe, an economist and educator, was a member of the Religious Society of the Quakers, and firmly believed in men's and women's equal rights to education. The establishment of TWCU was announced at the World Missionary Conference in Edinburgh in 1910.

1.5 Women's access to science education

In terms of women's access to science education, the TWHNS in 1905 opened research courses on top of the pre-existing four-year programme in three disciplines: humanities, sciences and home economics. However, there was only one place available. Since the primary goal of the Normal School was to cultivate female teachers, all the graduates had to fulfil the requirement to teach in schools. Thus, female students were unable to complete both the four years and the research programme ([5]: 185).

JWU also emphasised the importance of science education from the onset of its foundation. The founder, Jinzō Naruse, had studied at the Andover Theological Seminary and Clark University in Massachusetts from 1890 to 1894. In 1896, he published '*Jyoshi Kyōiku*' (Education for Women) in which he challenged the dominant view of that time that education for women does not benefit society. He proceeded to found JWU, which started with three faculties: home economics, Japanese literature, and English literature. He envisaged that the natural/social sciences were a component of the home economics curriculum. Distinguished professors such as Nagayoshi Nagai (pharmacologist, educationalist, first President of the Pharmaceutical Society of Japan) and Shōzaburō Watase were invited from the Tokyo Imperial University to give lectures on natural science subjects. Nagai's wife, who was from Germany, also taught the German language. In order to match the high standards of lectures given by those professors, in 1906, German-style laboratories and Leitz microscopes were installed in a new building designed for carrying out scientific experiments ([7]).

TWCU was also unique in that its second College President, Tetsu Yasui, created the Department of Mathematics in 1927, with the idea that 'the very thing that women need is mathematics'. In 1937, TWCU became the first women-only college that was accredited by the Ministry of Education, where the graduates were automatically granted the mathematics teacher qualification.

Even though some higher education institutions such as JWU existed for women at the turn of the century, it is worth noting that women had still not been allowed to become full-time registered students at any 'university' (*daigaku*), regardless of their specialisation.

1.6 Women's access to university in Japan

Women's access to the Imperial Universities (or any other publicly funded university) came thirty years after the establishment of the first Imperial University.

Figure 1.2 Autopsy class at JWU.

Source: Courtesy of Naruse Memorial Hall.

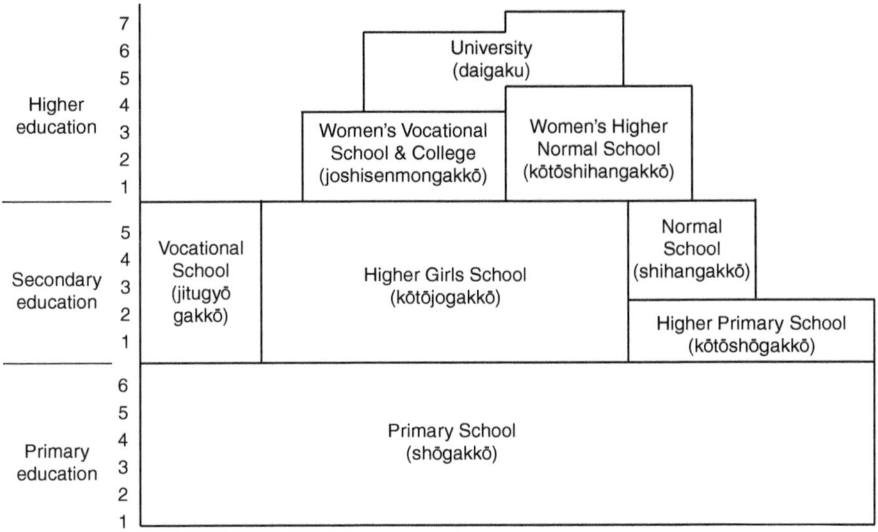

Figure 1.3 School system for women in pre-WWII Japan.

Source: [3] Adapted by the authors.

The year 1913 was the first year when female students were admitted to one of the Imperial Universities[4].

Yukawa (1994) examined this event at Tōhoku Imperial University in detail, and found that the first 'incident' of admitting a woman to the Imperial University was instigated by a decision of the then President of Tōhoku Imperial University, Masatarō Sawayanagi. Before taking the post as President, Sawayanagi was a civil servant in the Ministry of Education. His diary reveals that he held the firm idea that there is no need to create a separate, single-sex higher educational institution for women ([8]: 132).

His decision was met with severe criticism from the Ministry of Education, who sent a letter to him, enquiring about this 'serious incident that has no precedent' ([9]). Although there were no legal restrictions imposed on women to prevent them from entering universities, one of the prerequisites for university candidates was completion of high school, which meant in practice that no woman could be admitted to a university. Tōhoku Imperial University realised this requirement and thus included graduates of higher normal schools and those who were qualified as secondary school teachers, so that women could also enter. This rather radical change was only possible because of the strong commitment by President Sawayanagi as well as support from other distinguished male professors, including Tsuruichi Hayashi (Mathematics) and Rikō Majima (Chemistry), who had extensive experience in conducting research in Europe. The decision was also apparently influenced by the shortage of applicants for Tōhoku Imperial University (Tōhoku) at that time, compared with other Imperial Universities ([8]: 130).

Newspapers from that time also reflect how Japanese society in 1913 was not ready to accept the idea that women should or could study at university level as equals with men. The founder of JWU, Naruse, wrote that 'what we need to consider seriously is whether co-education is good or not'. Women's opinions on the issue were also divided. While a prominent female educator and founder of Kyōritsu Women's Vocational School (founded in 1886), Haruko Hatoyama, praised the courageous decision made by Tōhoku, another female educationalist, Ayako Tanahashi criticised it as 'women's debt to the nation is to raise their children' ([8]: 131).

Nevertheless, as a result of this change, four women (aged 24–40) applied and three were successfully admitted to Tōhoku. This is the first time in Japanese history that women were enrolled as full-time undergraduate students. Two of them, Chika Kuroda (Chemistry) and Raku Makita (Mathematics), were graduates of the TWHNS, and the third student, Ume Tange (Chemistry), was a graduate of JWU. Nagayoshi Nagai strongly supported the entrance of Kuroda and Tange. Subsequently, the two went on to conduct further research and completed their doctoral degrees in Science and Agriculture, respectively.

Kuroda was awarded her doctoral degree from Tōhoku in 1929, which made her the second female doctor in Japan after Kono Yasui (doctoral degree in Science, Tokyo Imperial University, 1927). Yasui completed her degree while working at the TWHNS. In contrast, Tange went on to the United States, and earned her PhD at Johns Hopkins University in 1927.

Interestingly, not even President Sawayanagi believed in the notion of women working outside the home as professionals after graduation. He apparently thought that only women from lower socio-economic classes should be allowed to participate in the unskilled labour force, and did not support women having professional careers.

1.4a Chika Kuroda

1.4b Raku Makita

1.4c Ume Tange

Figure 1.4 The first three female university students in 1913 (Courtesy of Tōhoku University and Naruse Memorial Hall, JWU), (a) Chika Kuroda, (b) Raku Makita and (c) Ume Tange.

Yukawa's study also demonstrated that after 1913 no other female student was admitted by Tōhoku in the following ten years. The archive shows that the government was clearly opposed to the idea, and Sawayanagi's decision was considered premature, at least before the liberal movements known as Taishō Democracy took off. It was in 1923 that Tōhoku and another private university, Dōshisha University in Kyoto, registered women as full-time undergraduate students. Between 1913 and 1923, women were allowed to become non-regular students or guest auditors at Tōyō University (1916), Hokkaidō Imperial University (1919), Kyoto Imperial University, Waseda University and Keiō University (1921). (See also Table 1.3)

Yukawa (1994) argues that women's right to education became more socially acceptable between 1913 and 1923. In particular, he emphasises the importance of the fact that Dōshisha University's decision to upgrade women's status as regular students was driven by the Christian principle of equal opportunities for men and women in education rather than individual conviction. It also sent a clear signal that a private university needed to provide women with an alternative to becoming teachers as the only career option available for women. The government's policy was solely focused on producing female teachers via the route of publicly funded higher education institutions (i.e. Women's Higher Normal Schools). Dōshisha University made a mark also by becoming the first private institution in Japan to opt for co-education ([10]).

After 1923, publicly funded universities such as the TWHNS and its all-men section in Hiroshima (Hiroshima Liberal Arts and Science University) followed the path of Tōhoku and Dōshisha. Furthermore, certain faculties in some universities came to terms with the idea of admitting women as full-time registered students. However, no demand existed in society for highly educated female professionals, outside a few select occupations such as doctors, nurses and teachers. The top two Imperial Universities (Tokyo and Kyoto) maintained their men-only policy until 1946, therefore not admitting women as regular undergraduate students until after WWII.

Table 1.1 reveals that Tōhoku accepted nearly half of all the female full-time university students prior to WWII, and Hokkaidō admitted one-third in those thirty years. In terms of colleges from which those women had graduated, twenty were from TWHNS, followed by JWU (11), TWCU (9) and Nara (8). It is interesting to see that out of those sixty-two, approximately half (30) specialised in mathematics, followed by fifteen in biology. Chemistry had nine, while physics had four. Half of the students were graduates from Higher Normal Schools in Tokyo and Nara, and their choices of subject were distributed across disciplines. Women's means of studying science were limited until the end of WWII and beyond.

1.7 Women's access to university in the West

It was therefore not widely accepted in Japan at the time that women should go to university, let alone contribute to academia and industry, despite the government's

Table 1.1 Undergraduate admission of women between 1913 and 1945 (By discipline/colleges they graduated from)

	Tōhoku	*Hokkaidō*	*Osaka*	*Kyūshū*	*Nagoya*	*Total*
Year Faculty of Science was founded	1911	1918	1931	1939	1942 (Separated from Engineering)	
Year women were allowed to enter	1913	1930	1931	1939	1942	
Year female admission was codified	1925	1930	1931	1939	1942	
Year of actual entry by women	1913	1930	1935	1942	1943	
Mathematics	10 8: TWHNS 1: Nara Women's Higher Normal School 1: JWU	10 4: TWCU 3: JWU 2: TWHNS 1: Qualified secondary teacher	5 3: TWCU 1: TWHNS 1: Nara Women's Higher Normal School (Temporary Teacher Training College)	1 1: Nara Women's Higher Normal School (Temporary Teacher Training College)	4 2: TWCU 2: Unknown	30
Chemistry	4 2: TWHNS 1: JWU 1: Beijing Normal University	2 1: JWU 1: Qualified secondary teacher	0	2 1: Imperial Women's College of Pharmacy 1: Kobe Women's College of Pharmacy	1 1: JWU	9
Physics	1 1: TWHNS	3 3: JWU	0	0	0	4

Biology	10 3: TWHNS 3: Nara Women's Higher Normal School 2: Imperial Women's College of Medicine and Pharmacy 2: Imperial Women's College of Science 1: Zhejian University	5 2: TWHNS 1: Nara Women's Higher Normal School 1: JWU 1: TWCU	0	0	0	15
Geophysics	1 1: TWHNS	0	0	0	0	1
Geology	0	3 1: TWHNS 1: Nara Women's Higher Normal School 1: Qualified secondary teacher	0	0	0	3
Total	26	23	5	3	5	62

(*Source*: [21]: 79.)

aspiration to 'catch up with the West'. In the United States, Oberlin College (Ohio) was founded in 1833 as a co-educational liberal arts institution, and in the United Kingdom, the University of London admitted women on equal terms with men in 1878. Two female students were awarded Bachelor of Science degrees in 1881. In France, the Faculty of Medicine in Paris, admitted the first female students in 1867. In the late 19th century, Paris and other European cities such as Zurich attracted foreign female students who wanted to study medicine and other subjects ([11] [12] [13]).

In 1875, universities became open to women in Denmark, followed by Italy, the Netherlands, and the UK (1876). Australia, Canada, and France followed suit in 1880. In Belgium, women were successively admitted to the Universities of Brussels, Liège and Ghent between 1880 and 1882, although the Catholic University of Louvain did not open its doors to women until 1920 ([14]). In Ireland, the first nine women graduated from the Royal University of Ireland in 1884 (it was subsequently separated into the National University of Ireland and Queen's University Belfast in 1909), while the University of Dublin (Trinity College Dublin) only accepted female students in 1904 ([15]). In Spain, women were only allowed to enter for private university degrees by special dispensation, until universities became fully open to them in 1910. In Germany, Heidelberg University and Freiburg University first admitted women as regular students in 1890, and a ministerial decree of 18 August 1908 legalised the status of regular women students in all universities in Prussia. At the time when only three women were admitted to a university in Japan, over 20 per cent of university students were women in the UK, 16 per cent in Finland, 10 per cent in France, 8 in Austria, 6–7 per cent in Germany and Italy and 3 per cent in Belgium ([13]:189).

In Japan's neighbouring country, China, Peking University opened its door to women in 1920, followed by other universities nationwide. Korea's first women's educational institution was founded in 1886 by the American Methodist Episcopal missionary Mary Scranton. It was upgraded to university status in 1910 (and later became Ewha Women's University).

Compared with these global trends, Japan was very slow in accepting women as regular students at university level. Therefore, the individual convictions of a few (predominantly male) academics, administrators and individual female researchers played a crucial role in increasing female participation in higher education and scientific research. However, the restricted access to higher education was not the only obstacle to Japanese women in STEM. They also faced difficulties in entering research professions.

1.8 Demands for scientific research and skills

The nascent form of the national science institution developed at around the same time. Rikagaku Kenkyūjo (RIKEN, the Institute of Physical and Chemical Research) was originally proposed by renowned chemist Jōkichi Takamine in 1913, and established in 1917, after the Diet passed the relevant resolution. RIKEN was funded jointly by the government, industry and the Imperial Household, and later became a publicly funded, national science research institute in 1958.

RIKEN played a significant role for some female scientists. When women were not allowed to study as equals of men at Imperial Universities, RIKEN provided female researchers with a place for conducting research alongside their daily jobs as teachers and lecturers at the TWHNS or women's colleges.

Female scientists with doctoral degrees, including Chika Kuroda, Ume Tange, and Michiyo Tsujimura (doctoral degree in Agriculture, Tokyo Imperial University, 1932) continued working at RIKEN. Others had the opportunity to continue their research activities at various Imperial Universities, while simultaneously fulfilling their duties as teachers.

According to Nagashima's book published in 1937, there were 12,356 Japanese scholars who had doctoral degrees, of whom only 21 were women, as of April 1937. Two-thirds of these (14 out of 21) were awarded a doctoral degree in medicine. Ten years later, another ten women achieved the highest level degree, bringing the total to 31 (14 in Medicine, 11 in Science, 4 in Agriculture and 2 in Pharmacy) ([5]: 191-192). Out of these 31 female scientists, 14 had been awarded doctoral degrees in medicine, eleven of whom had studied at the Tokyo Women's Medical University, and 11 out of 17 women with degrees in other disciplines attended Women's Higher Normal Schools (either in Tokyo or Nara). Their doctoral degrees were awarded from six Imperial Universities in Tokyo (9), Kyoto (8), Tōhoku (6), Hokkaidō (2), Osaka (2) and Kyūshū (1) as well as one private university Keiō (3).

The first and second Japanese female scientists who obtained doctorates abroad were Tsuruko Haraguchi (1886–1915, Ph.D. in 1912) and Tomi Kōra (1896– 1993, PhD. in 1922). They were both JWU graduates, and went to Columbia University to complete their doctorates in psychology ([16]).

Notably, there is a lack of doctoral research in the fields of physics and engineering. The first female student in physics was Toshiko Yuasa (1909–1980). Yuasa entered the Department of Physics, Tokyo Liberal Arts and Science University (predecessor of Tokyo University of Education, presently Tsukuba University) in 1931, as the first Japanese woman to specialise in physics. At that time, Tokyo Liberal Arts and Science University was the only university located in the Kantō (Greater Tokyo) area among five universities across the country that accepted female regular students (Tōhoku, Hokkaidō, Kyūshū, two Liberal Arts and Science Universities in Tokyo and Hiroshima). She wished to go to Tōhoku, but her mother opposed the idea.

Inspired by the discovery of artificial radioactivity in France, Yuasa decided to go there. Her decision was supported by her father who was an engineer. The environment for experimental physics in Japan at that time was far from ideal, with a shortage of equipment and staff, and particularly difficult for a female associate professor at TWHNS. In 1939 Yuasa passed the examination given by the French government and went to Paris in 1940, even though WWII had just started in Europe. Professor Jean Frédéric Joliot-Curie who discovered artificial radioactivity, became her advisor at the Collège de France.

Later, Yuasa received a French degree in physics (Doctorat en Science) for her research on the beta-ray spectra from artificial radioactive nuclei in 1943. Upon her return from France, she became a professor at the TWHNS (1945 to early

Table 1.2 Female scientists awarded doctoral degrees between 1927 and 1940 (excluding medicine)

Name	*Undergraduate degree-awarding institution*	*Date of PhD award*	*PhD awarding institution*	*Degree*
Kono Hosoi	TWHNS	20/4/27	Tokyo	Science (Botany)
Chika Kuroda	TWHNS & Tōhoku	4/11/29	Tohoku	Science (Chemistry)
Sechi Kato	TWHNS & Tōhoku	8/6/31	Kyoto	Science (Chemistry)
Michiyo Tsujimura	TWHNS	8/6/32	Tokyo	Agriculture
Yasu Honma	TWHNS	4/5/36	Hokkaidō	Agriculture
Hideru Suzuki	JWU	12/1/37	Tokyo	Pharmacy
Fumiyo Ogawa	Nara Women's HNS & Tōhoku	5/4/38	Tōhoku	Science (Biology)
Yasu Hatagoshi	Nara Women's HNS	4/5/39	Kyoto	Agriculture
Shizuko Matsumoto	Osaka Prefectural Women's College	9/10/40	Kyoto	Science (Biochemistry)
Ume Tange	JWU & Tōhoku	14/8/40	Tokyo (& Johns Hopkins, USA, 1927)	Agriculture

(*Source*: [5]: 193.)

1949), but once again found her research environment to be inadequate. The Occupation Forces in Japan prohibited any research in experimental nuclear physics. She eventually decided to leave Japan, relying on her mentor Professor Joliot's help. In 1957 she became Maître de Recherche (chief researcher) at CNRS, and continued her research at l'Institut de Physique Nucleaire de l'Université de Paris, Orsay, where the former institute had moved between 1957 and 1959. After 1949, she remained in Paris until her death in 1980 ([17]).

In the inter-war period, Japan faced a number of challenges, one of which was a shortage of educated and skilled men. This prompted the development of initiatives to encourage educational institutions to produce graduates with numerical and scientific skills, regardless of gender.

In the late 1920s, several science courses at several universities were established for women. Major initiatives included the Imperial Women's Medical College (1925), Nitobe's brainchild, the Women's College of Economics (1928), Imperial Women's College of Science (1941) and Kyoto Prefectural University Women's College (1942). In 1943, Tsuda College created 30 places each for physics, chemistry, and mathematics, while more women's colleges started three-year programmes from 1944 onwards. However, at the end of WWII, a new era arrived for higher education in Japan. As of 1941, the number of female regular students at various universities in Japan is summarised in the table below.

Table 1.3 Total number of female students (registered/non-registered) at Imperial and private universities in 1941

	Natural Sciences	*Humanities*	*Liberal Arts & Science*		*Graduate School*	*Elective/Special/ non-regular*	*Non-degree graduate course*	*Total*
			Arts	*Sci.*				
Tokyo Imperial University					1			1
Kyoto Imperial University							7	7
Tōhoku Imperial University		4			2	1		7
Kyūshū Imperial University		2				1	2	5
Hokkaidō Imperial University	6						1	7
Osaka Imperial University	2				1		4	7
Nagoya Imperial University							2	2
Niigata Medical U. (–1960)							1	1
Chiba Medical U. (–1960)							4	4
Kanazawa Medical U. (–1960)							1	1
Tokyo Lib. Arts & Science U.			2	2		21		25
Hiroshima Lib. Arts & Science U.			4	2		1		7
Public (Total)	**8**	**6**	**6**	**4**	**4**	**24**	**22**	**74**
Keiō						1		1
Waseda		9				8		17
Meiji		20						20
Hōsei						3		3
Nihon						44		44
Dōshisha		19				1		20
Tōyō		5				4		9
Private (Total)	**0**	**53**	**0**	**0**	**0**	**61**	**0**	**114**
Public + Private (Total)	**8**	**59**	**6**	**4**	**4**	**85**	**22**	**188**

(*Source*:[5]: 189.)

1.9 Women and gender equality in Japanese society before WWII

In the Meiji era, the idea of gender equality was imported from the West to Japan by some scholars. Yukichi Fukuzawa, the founder of a private university and its affiliated primary and secondary schools, Keiō, was one of the early proponents of gender equality. He opened a private primary school with one of his pupils, Yoshirō Wada, in 1874 and for a few years the school was co-educational ([19]). However, it met with fierce opposition, particularly from former warrior families, and the school reverted to its earlier gender-segregated educational system.

In the 1880s, a people's rights movement was led by several prominent female figures such as Hideko Fukuda and Toshiko Kishida. They are regarded as the first group of feminists in Japan. Both of them were highly educated for that time and were also involved in women's education. Some of the reforms around this period include: the Emancipation Decree for Female Performers (*geisha*) and Prostitutes (Geishōgi Kaihō Rei, 1972), the granting to women of the right to divorce their husbands (1873), the establishment of vocational schools for women (1873) as well as that of women's normal schools (1874). Another movement led by a prominent woman, Kajiko Yajima, was the Women's Kyōfūkai Association, which promoted temperance as the Japanese branch of the Women's Christian Temperance Union. Yajima was also an educator, and later became the first president of a private all-girls' school, *Joshigakuin*, in 1890.

However, their movements were not widely supported due to strong resistance from both men and women of different classes. One notable example of the backlash was the legal ban on any political activity by women in 1890. Article 5 of the legislation banned women, along with military men, policemen, Shintō priests and teachers, from forming political associations, and denied these groups freedom of assembly, which included joining or hosting political meetings.

In terms of suffrage, Japan achieved local women's suffrage at an early stage for international standards, yet only for a very short period of time. In 1878, in Kōchi Prefecture, Kita Kusunose, widow and head of a household, made an official complaint that it was unfair and unlawful for her as a taxpayer to be denied suffrage, just because she was a woman. The prefectural government denied her the right, but she brought the case forward to the Ministry of Internal Affairs. They issued the Order Regarding Ward, Town and Village Assemblies in 1880, and granted suffrage to female heads of household. However, it was not long before the government amended the Order in 1884, dropping this clause, and women lost the right to vote. Political rights movements began gathering speed in the early twentieth century. The Taishō era (1912–1926) became associated with liberal movements (known as the Taishō Democracy movements), with its universal suffrage movement culminating in the General Election Law enacted in 1925. The law extended suffrage to all males aged 25 and over but females were still not granted the right to vote.

The period was the real beginning of feminist movements in Japan. An association called *Seitō-sha* (Bluestocking Society) and its journal *Seitō* (Bluestocking) were founded in 1911. Several JWU graduates as well as a feminist and peace campaigner, Akiko Yosano, were involved in the launch, but the principal architect was Raichō Hiratsuka. Women's suffrage was on the agenda, and in 1920 Hiratsuka also founded the New Women's Association (*Shin Fujin Kyōkai*). Their activities were highly controversial, even in the context of the rising liberalism of the time. Both male and female educationalists, such as Jinzō Naruse and Umeko Tsuda, neither supported their activities nor accepted their criticisms of gender inequality. However, it was the nascent period of feminist movements in Japan, and put questions concerning gender equality to wider society for the very first time.

Between 1918 and 1919 there was a dispute amongst feminists, which later became known as the 'Protection of Motherhood Debate'. Hiratsuka demanded that the government should cover all the costs of pregnancy and delivery, whilst Akiko Yosano criticised this by claiming that maternity-centrism was merely an alternative way of supporting the concept of 'good wives, wise mothers'. Waka Yamada refuted both by stating 'women should not depend on men or the nation' from the standpoint of being 'good wives, wise mothers'.

Hiratsuka went on to form the first women's political association, the New Women's Association, together with Fusae Ichikawa and Mumeo Oku. Ichikawa was opposed to the idea of 'good wives, wise mothers' altogether. Waka Yamada was also a founding member, despite her earlier dispute with Hiratsuka over motherhood and the role of government. Although its momentum declined due to internal divisions and external pressure, their association brought about some remarkable outcomes in acquiring political and social rights for women. These include the revision of Article Five of the Security Police Act (1922), which banned women from joining political assemblies, and the amendment of the Lawyers Act (1933), which enabled women to become lawyers.

The Japan Women's Suffrage Association was also founded in 1921 (later renamed the Japan Woman's Christian Suffrage Association) by Tsuneko Gauntlett and Ochimi Kubushirō. However, as previously discussed, women were not granted the right to vote by the 1925 General Election Law. Fusae Ichikawa started Japan's first women's suffrage organisation, the Women's Suffrage League of Japan (*Nippon fujin yūkensha dōmei*) in 1924, and campaigned for the cause. She went to the US and worked for the International Labour Organisation (ILO). In 1930, the League instigated Japan's first national convention on the enfranchisement of women. Although women's suffrage was only achieved after WWII in 1945, the co-founder of the Women's Suffrage League, Shigeri Yamataka, became one of the rare female politicians in the Upper House after WWII (1962–1971). Ichikawa herself served the Upper House for five terms (25 years in total) from 1953. Ichikawa and Yamataka were both educated in Women's Normal Schools (in Aichi prefecture and Tokyo, respectively). The reluctance to embrace 'modernisation' meant much slower progress in granting women equal opportunities in society, at least according to

Western standards. However, some outcomes deriving from women's access to higher education began to be seen during the inter-war period.

1.10 Conclusion

When the modernisation of Japan was on the agenda, there was a clear mission for the government to catch up with the West in many fields, including education. The first element, necessary for change (pressure for internationalisation) was evident and the government swiftly created a school system modelled upon that of the United States to fulfil the ultimate goal of modernising the country's economy and industry. However, the government wanted to maintain some traditional social norms and values while trying to catch up with the West. Thus a difficult balance had to be struck between educating women and keeping them inside their homes as much as possible. The national policy clearly steered away from treating women as the equals of men in both society and the education system, but some individuals strove for greater equality.

A lot of Western thinking was imported into Japan during this period, which led to the creation of higher education institutions for women. The governmental policy of developing female teachers through the (Higher) Normal Schools, combined with some privately funded colleges for women, secured the provision of an environment where a few women could strive to pursue their dreams of becoming scientists.

The history of women in science in Japan in these times exemplifies the interaction between individuals and the environments they found themselves in. Apart from several individuals, formal institutions, including academia and the government, were unsupportive of women who tried to pursue any professional career, let alone one in science and engineering.

Some of the graduates from Normal Schools or private colleges for women became suffragettes advocating social reform. The second element (social awareness of gender inequality) was also emerging during the early twentieth century. Yet the divisions amongst women themselves ran deep, and both male and female educators campaigning for women disagreed with the goals of some feminist groups. As a result, no social consensus existed for providing equal opportunities for women to study at the highest educational institutions (i.e. Imperial Universities), let alone specialising in STEM. Although some individuals within the academic community (e.g. at Tōhoku Imperial University) took the initiative to open the door to women, there was no coordinated effort at institutional level. The third element (unity of the academic community towards gender equality) was particularly lacking in this era. Free access to university education and suffrage were both finally achieved only after the end of WWII.

Notes

1 These laws included the Middle School Order (1886), Normal School Order (1886), Vocational School Order (1899), Middle School Order Amendment (1899), Vocational

School and College Order (1903), Primary School Order Amendment (1907), High School Order (1918) and University Order (1918).

2 The other six Imperial Universities were: Kyoto Imperial University (1897–1947), Tōhoku Imperial University (1907–1947), Kyūshū Imperial University (1911–1947), Hokkaidō Imperial University (1907–1947), Osaka Imperial University (1931–1947), and Nagoya Imperial University (1939-1947). Keijō Imperial University (1924–1946) was also founded during Japan's imperial expansion period. In the post-WWII years, it was transformed into the Seoul National University. Similarly to Keijō, Taihoku Imperial University became the National Taiwan University after WWII.

3 The first licensed female physician was Ginko Ogino (who was the only woman graduate from a private medical academy, now Juntendō University). After experiencing hardship and difficulties, she was finally allowed to take the medical practitioner's examination, becoming the first registered female doctor in 1885.

4 Subsequently, Kyūshū Imperial University admitted female students in 1925, followed by Hokkaidō Imperial University and Osaka Imperial University in 1935.

References

[1] E. Westney. *Imitation and Innovation: The Transfer of Western Organizational Patterns to Meiji Japan*. Cambridge, MA: Harvard University Press, 1987.

[2] M. Kanda. *Nihon no shiten* (A Japanese perspective): Talk prepared for Japan Education Forum II. [In Japanese], 2005.

[3] Ministry of Education, Culture, Sports, Science and Technology. *Gakkō Keitōzu* (Diagram of School System) [In Japanese]. [online]. Available from: http://www.mext.go/b_menu/hakusho/html/others/detail/1318188.htm. Accessed 14 December 2014.

[4] Ministry of Education, Culture, Sports, Science and Technology. The increase of government normal schools. [online]. Available from: http://www.mext.go.jp/b_menu/hakusho/html/others/detail/1317284.htm. Accessed 15 September 2013.

[5] A. Tsugawa and Y. Kanomi. *Hiraku: Nihon no josei kagakusha no kiseki* (Opening: Tracing the paths of Japanese female scientists). [In Japanese] Domesu Publishing, 1996.

[6] S. Yoshimi. *Daigaku towa nanika* (What is a university?). [In Japanese] Tokyo: Iwanami Publishing, 2011.

[7] Japan Women's University. *Saiensu ni idomu jyosei tachi* (Women in science who take on the challenge). *[In Japanese]*. Tokyo: Ad Three Publishing, 2009.

[8] T. Yukawa. *Taishō-ki ni okeru jyosei no daigaku no monko kaihō: Taishō 2 (1913) nen no Tōhoku Teikoku Daigaku no jirei to sonogo no tenkai* (Admitting women to universities in Taishō Era: the case of Tōhoku Imperial University in 1913 and its aftermath). [In Japanese] *Kyōikugaku Kenkyū (Japanese Journal of Educational Research)*, 61(2): 129–138, 1994.

[9] Y. Motomura. *Dokusō no keifu: jyoshi gakusei, teikoku daigaku nyūgaku 100nen, uketsugareru tsuyoi ishi* (Genealogy of originality: 100th anniversary of admitting female students to Tōhoku Imperial University, passing down strong will and determination). [In Japanese]. *Mainichi Shinbun*, 1 August 2013.

[10] Doshisha University. *Joshi kōto kyōiku no shidō* (Start-up of higher education for women [In Japanese]. [online]. Available from: http://www.dwc.doshisha.ac.jp/about/records/publication/125_years/chapter2_1.html. Accessed 20 September 2013.

[11] C. Schultze. *Les femmes médecins au XIXe siècle* (Women Doctors in the 19th Century). [In French] Paris: Ollier-Henry, 1888.

[12] L.L. Clark. *Women and Achievement in Nineteenth-Century Europe*. Cambridge: Cambridge University Press, 2008.

[13] N. Tikhonov. *La mixité dans l'éducation. Enjeux passés et présents*, chapter Enseignement supérieur et mixité: la Suisse, une avant-garde ambiguë. (Education and diversity: Switzerland, an ambiguous vanguard.) [In French], pages 35–52. Lyon: ENS Editions, 2004. ed. R. Rogers.

[14] N. Tikhonov. Student migrations and the feminisation of european universities (summary). actes de l'histoire de l'immigration (electronic journal). [online] Available from: http://barthes.ens.fr/clio/revues/AHI/articles/english/tiko.html, 2002. Accessed 15 November 2014.

[15] J. Harford. *Have Women Made a Difference? Women in Irish Universities, 1850–2010*, chapter 'Women and the Irish university question', pages 7–28. Bern: Peter Lang, 2010. eds. J. Harford and C. Rush.

[16] M. Takasuna. Nihon jyosei shinrigakusha no hakasego (the first Japanese female psychologist with a PhD. [In Japanese]. [online]. Available from: http://www.psych.or.jp/publication/worold_pdf/58/58-2-3.pdf, 2012. Accessed 15 September 2013.

[17] E. Yagi and H. Matsuda. Toshiko Yuasa (1909–80): the first Japanese woman physicist and her followers in Japan. *AAPPS Bulletin* (the Association of Asia Pacific Physical Societies), 17(4): 15–17, 2007.

[18] Y. Fukuzawa. Nakatsu ryūbetsu no sho (a letter of farewell to Nakatsu) [In Japanese]. [online]. Available from: http://www.aozora.gr.jp/cards/000296/card46890.html, 1872. Accessed 3 September 2014.

[19] S. Sievers. *Flowers in salt: the beginning of a feminist consciousness in modern Japan*. Stanford: Stanford University Press, 1983.

[20] S. Sekiguchi. Confucian morals and the making of a 'good wife and wise mother': From 'between husband and wife there is a distinction' to 'as husbands and wives be harmonious'. *Social Science Japan Journal*, 13(1): 95–113, 2010.

[21] M. Yamamoto. Hokkaidō Teikoku Daigaku e shingakushita Tokyo Jyoshi Daigakusei tachi (Tokyo Woman's Christian University Students who went on to study at Hokkaidō Imperial University) [In Japanese]. *Hokkaidō Daigaku Bungaku Shokan Nenpō (Annual Report of Hokkaido University Archives)* 5: 76–103, 2010.

2 Policy changes from the post-WWII period to the mid-1980s

2.1 Introduction

This chapter first outlines the higher educational landscape in post-WWII Japan, highlighting the government's move towards a co-educational system, and the widening access of women to higher education, with the enactment of the Fundamental Law of Education and the School Education Law in 1947. It then looks at policy developments around women's rights in general, with a particular emphasis on the EEOL in 1986. Both these government initiatives were taken rather passively in response to international pressure. However they served as catalysts for further changes for higher education institutions, and paved the way for future policy developments. As discussed in Chapter 1, this chapter demonstrates that two factors were important, namely 'pressure to westernise' and 'social trends and campaigns regarding gender equality'.

The chapter also describes the significant problem of the underrepresentation of women in STEM into the 1990s, and the activities of academic societies in scientific fields during that period. The most significant factor in bringing about government policy change in the early 2000s was the academic community's active involvement in presenting the facts about the underrepresentation of women in STEM to the wider society; that is, projecting the problem in line with the government agenda of equal participation of men and women. The ultimate priority of the government at this time was to increase the historically low fertility rates and declining population numbers.

2.2 The higher educational landscape in post-WWII Japan

After WWII, the General Headquarters (GHQ) of the Allied Forces requested a report from the US Educational Mission for recommendations concerning educational reform in Japan. Following the report, the Educational Renewal Committee was established under the direct supervision of the Cabinet in August 1946. This report and subsequent deliberations laid the foundation for the post-war educational policy in Japan.

The Educational Renewal Committee (later renamed Educational Renewal Council) submitted 'a total of 35 proposals to the Cabinet, from the first proposal concerning the Fundamental Law of Education and the school system in December 1946 to the last proposal for the establishment of the Central Council for Education in November 1951. Almost every law, ordinance, and system related to educational reform in postwar Japan was enforced after discussions at the committee/council. In that sense, it played an extremely significant role in post-war educational reform in Japan' ([1]).

In 1947, the new university system was introduced by the Fundamental Law of Education and the School Education Law, under which the school system was overhauled. Japan adopted the '6-3-3-4' school system: six years of primary school, three years of lower secondary school, three years of upper secondary school, and four years of university. Unlike the pre-WWII educational system, it became a single-track and uniform system. As a result, various secondary educational institutions were reorganised and integrated into lower secondary (junior high) schools for three years and upper secondary (high) schools for another three years. In tandem with the new school system, the period of compulsory education was extended to nine years, including six years at primary school level plus three years at lower secondary school level. All children were educated in a fundamentally standardised school system. The recommendations from the Education Renewal Council contained proposals for universal access to education and co-education.

In higher education, women were finally granted the option to attend university in 1946 when the University of Tokyo and Kyoto University opened their doors to

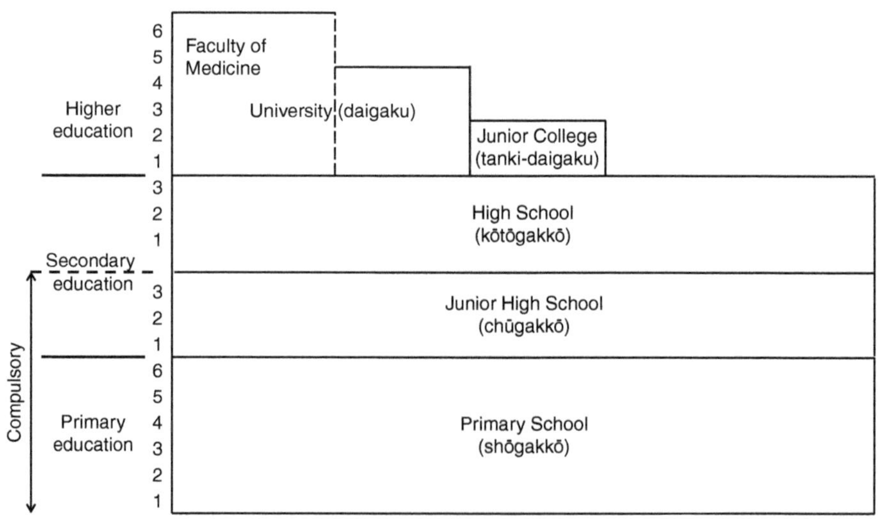

Figure 2.1 School system in post-WWII Japan (as of 1949).

Source: [18]. Adapted by the authors.

female students ([16], [17]). Furthermore, women were admitted to various national research institutes such as forestry and fisheries ([2]: 191).

In April 1948, prior to the nationwide launch of the reformed university system, 11 private colleges and one prefectural college were upgraded to university status. They were Kokugakuin, Sophia, JWU, TWCU, Tsuda College, Sacred Heart, Dōshisha, Ritsumeikan, Kansai, Kwansei Gakuin, Kobe College and Kobe University of Commerce. Five of them (JWU, TWCU, Tsuda, Kobe College and Sacred Heart) were women-only colleges, and seven out of these 11 were established based on Christian principles ([19]). As the upgrading of women's colleges and higher normal schools to university status had been considered for nearly two decades, the reform was seen as a milestone by the stakeholders ([3]).

In the following year, TWHNS and Nara Women's Higher Normal School became the only two national women-only universities, as Ochanomizu University and Nara Women's University, respectively. The three women with doctoral degrees (Kono Yasui, Chika Kuroda, and Toshiko Yuasa) were promoted to professor level in the Faculty of Science at Ochanomizu University. At the same time, the Imperial Women's College of Science became co-educational and was renamed Tōhō University in 1950. Tsuda College dropped its physics and chemistry courses, but retained mathematics. Additionally, in the late 1950s, JWU had a plan to reorganise the Faculty of Home Economics into the Faculty of Science (which did not materialise until 1992).

As previous chapters described, a secondary-level school teaching job was almost the only option available for women who had majored in science and engineering in the past. Therefore, if one had an aspiration to become a scientist or an engineer, the only way to achieve this goal was to be educated at either one of the two National Women's Universities or a private women's college, and then continue postgraduate study while working as a teacher. The choice was often made at the expense of marriage and family. There was certainly no guarantee of obtaining a post as a scientist after receiving the postgraduate degree; the risk was simply too high.

After World War Two, although co-ed became the basic education policy of the country, almost 30 years passed before women began to take the opportunity to study at a co-educational national university. Even then, becoming a scientist or working along with men was an exception.

First of all, the ratio of women progressing to higher education did not increase immediately after the war. Second, single-sex education schooling remained a popular choice at secondary education level and above all among parents. The competitive entrance examination system used for elite universities reinforced the trend that good students were attracted to single-sex schools. Single-sex high-schools have been more successful in attaining better outcomes in the single-track uniform system. All-girls' high schools such as Ōin Gakuen and Joshigakuin gained a reputation as good educational institutions in the early 1970s. Yet only a tiny fraction of their very top pupils decided to go to the University of Tokyo (fewer than 10 students). This trend continued until the 1990s. ([4])

At the higher education level, opening access to women at the end of the war did not mean that institutions were ready to accommodate more female students. Women had a choice between four-year women's universities on the one hand and two-year junior college (tanki daigaku, tandai) on the other. Large national universities continued to be male-dominated. Successful rates of being hired after graduation from a two-year liberal arts college has long been very high for women, which created the cycle of parents encouraging their daughters to enter those colleges and firms hiring them for a few years before they find their husbands and quit their jobs. This cycle evidently delayed women's professional progression, particularly in science and technology.

Co-education became the official line of government educational policy after World War Two. In many cases, Higher Schools for girls became co-ed institutions. Furitsu Daiichi Kōjo (Tokyo Metropolitan Hakuo Senior High School) and Kagoshima Daiichi Kōjo (now Tsurumaru High School) are examples. However, in some cases, particularly, in the Kantō region, Urawa Daiichi Kōjo (Urawa Daiichi Girls' High School), Utsunomiya Daiichi Kōjo (Utsunomiya Girls' High School) remained single-sex. In Chiba, one school became co-ed, but later changed back to being a girls' school. In the past, as secondary schools were mostly single-sex, women had limited access to high schools. Therefore, going co-ed and opening access to higher education for women almost at the same time was a really radical step in Japan.

While single-sex education for girls became a relic of the past in the public education sector, private girls' schools remained single-sex in many cases. In particular, private girls' junior and senior high schools attracted certain types of academically gifted students since the 1980s. Some of the well-reputed girls' schools include Joshigakuin, which started as Japan's first mission school in 1870, Kobe Ehwa Girls' School (1875), Doshisha Girls' School (1878), Japan Women's University High School (1901), Futaba Gakuen (1909), and Sacred Heart Girls' High School (1910). It is worth noting that almost all of these schools were established as Christian foundations ([5]: 44).

In addition, the private girls' high school, Ōin Gakuen, is well known and historic. The school was founded in 1924 by graduates of the then Tokyo Women's Higher Normal School (Ochanomizu University). Another example is Keiō Girls' High School, established in 1950 as an affiliated secondary school to Keiō University.

The high school enrolment rate has continued to rise during the period of rapid economic growth. It reached more than 90 per cent in the mid-1970s, and then plateaued. Initially, girls' enrolment rate was low, but since the 1970s, it exceeded the male enrolment rate, and is currently almost equal to the male enrolment rate.

In spite of these new initiatives in higher education, women entering universities and two-year colleges accounted for only 1.2 per cent of the total eligible population in 1950. Their numbers increased over time, particularly from the 1960s onwards (see Figure 2.2).

After the war, the admission rate for women entering universities and two-year colleges rose rapidly, particularly in the 1960s. When the two figures are combined,

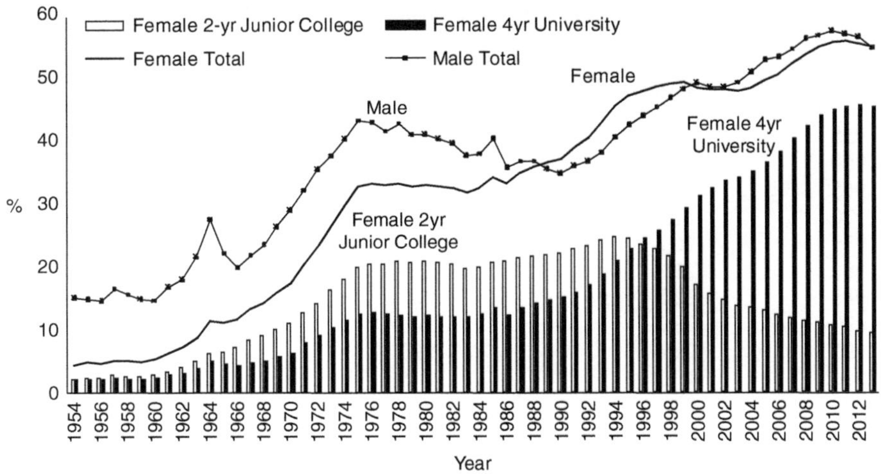

Figure 2.2 Tertiary education participation rates, 1954–2013.

Source: [20], various years. Adapted by the authors.

they exceed 10 per cent. In the 1970s, with this trend continuing, it rose to more than 20 per cent. In fact, in the early 1960s, a professor in a private university wrote in a magazine column that a further increase in the number of female university students would eventually ruin the country ([6]: 18).

However, women's entry to the former seven Imperial Universities at undergraduate level only increased recently. Even in 2009, the ratios at most of these seven universities (19.0 per cent for the University of Tokyo and 21.6 per cent for Kyoto University[1]) were still remarkably low, compared with the average ratio of female students (40.7 per cent) for all (national, local and private) types of universities.

Furthermore, the government reinforced traditional gender roles ('men work outside, while women stay at home') through the national curriculum, in response to the high demand for rapid economic growth. In 1958, the Ministry of Education introduced a discriminatory system of gender-tracking at secondary schools. While boys took technology courses, girls studied home economics ([7]).

A suffragette and politician, Fusae Ichikawa (1893–1981), launched a campaign against this gender discriminatory policy in 1974. However, the policy continued until 1989, when the Ministry of Education finally made home economics a co-educational course, mainly in response to the above-mentioned UN Convention on discrimination against women ([8]). The general tone of the national curriculum meant that sex discrimination remained normative, and single-sex education at secondary and tertiary levels was widely accepted until the 1990s.

Despite a short period of strong feminist movements after WWII, women's universities (including two-year colleges) remained the most desirable choice for many wealthy and middle-class families for a number of years, as alternative options were viewed as overly risky for their daughters' future role in society.

Table 2.1 Enrolment rate for women in four-year universities and two-year colleges, 1955–2010

Year	*2-year college (%)*	*4-year university (%)*	*Total (%)*
1955	2.6	2.4	5
1960	3	2.5	5.5
1965	6.7	4.6	11.3
1970	11.2	6.5	17.7
1975	20.2	12.7	32.9
1980	21	12.3	33.3
1985	20.8	13.7	34.5
1990	22.2	15.2	37.4
1995	24.6	22.9	47.5
2000	17.2	31.5	48.7
2005	13	36.8	49.8
2009	11.1	44.2	55.3

(*Source*: [20], various years. Adapted by the authors.)

Table 2.1 shows how many 2-year colleges and universities were built as the number of entrants to higher education increased after the 1960s.

In two-year junior college, the enrolment rate among women has almost doubled every five years from 1960 to 1975. Although the enrolment rate for women in a four-year university slowed down between 1965 and 1970, it exceeded 10 per cent over the next five years (1970–1975) and stabilised until 1990.

The success rate of obtaining a job was 57.4 per cent for graduates from a two-year college in 1965. This was the first time that more than 50 per cent of two-year college graduates entered the labour market. In the 1970s, the employment rate for two-year college graduates was 68.8 per cent, which was greater than 59.9 per cent for university graduates. It is safe to assume that companies preferred hiring women from a two-year college as *shokuba no hana* ('office flowers', i.e. beauty in the office).

Higher education for women became widely accepted as important, and their participation level began to increase. However, there remained a two-track education system, one being a four-year university and the other being a two-year college. There were also many women-only universities, as previously mentioned. A genuine equality in co-educational terms was only achieved from the mid-1990s onwards, with the EEOL enacted in 1986 being a milestone. This will be described in more detail later.

The next sections will look at the formation of an association of female scientists in the 1950s, also the establishment of a new educational institution (vocational high school) in response to calls for more technical training and education for the fast-growing economy in the 1960s up to the end of 1980s.

2.3 Into the 1960s: founding the Society of Japanese Women Scientists

Despite the small number of female scientists and the lack of demand for professional women researchers, a group of female scientists in Japan formed the

Society of Japanese Women Scientists (SJWS) in April of 1958. Its establishment was instigated by an external force, and in direct response to an international peace movement.

The Committee of Seven to Appeal for World Peace was a non-partisan group of seven members. It was initially formed in 1955 by seven prominent intellectuals including the Nobel Laureate (Physics) Hideki Yukawa and feminist campaigner Raichō Hiratsuka. The Committee urged a small number of female scientists in Japan to select and send delegates to the 4th World Women's Conference in Vienna in 1958. At the time, Hiratsuka was the President of the Federation of Japanese Women's Organizations and the Vice President of the Women's International Democratic Federation. She believed that Japanese female scientists had a duty to contribute to world peace.

The Society was founded in 1958, which was six years before the first International Conference of Women Scientists and Engineers (ICWES) was organised in New York in 1964. SJWS has been a regular participant in the ICWES conferences since its inception.

For the first nine years, there was no President for SJWS. Being the first female student admitted to a national Imperial University in 1913 and an eminent chemist, Chika Kuroda (1884–1968, Ochanomizu) was approached to take the post but declined, and became an honorary advisor, while also acting as President for nearly ten years.

The first SJWS President (1969–1979) was a chemist, Kimiko Anno (1910–2009, Tokyo Liberal Arts and Science University). She was the first graduate student admitted to Tokyo Imperial University. Apart from these findings, the fact remains that women in science as a whole were still very much a minority group. Because of its minority status, the group did not have much direct influence in

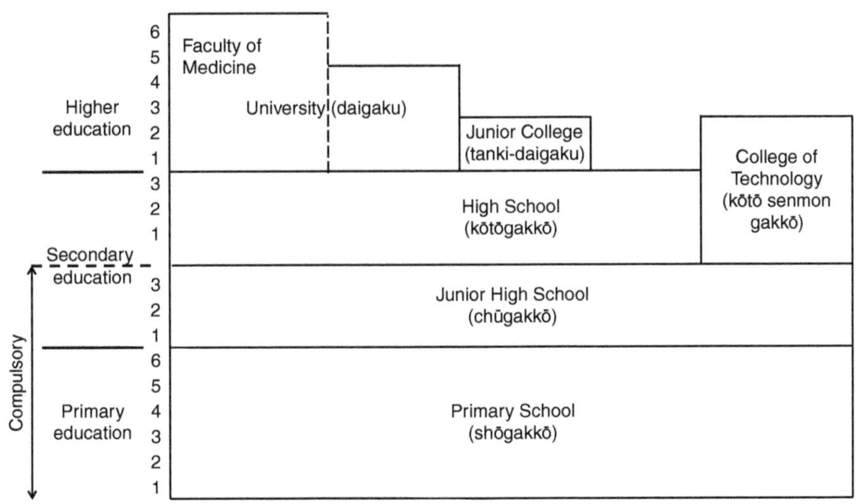

Figure 2.3 School system in post-WWII Japan (as of 1972).

Source: [18]. Adapted by the authors.

increasing the number of female scientists for a long time. However, when the underrepresentation of female scientists remained unresolved in the 1990s, the SJWS raised concerns in various fora in academia and the policymaking community (Source: personal communication).

By the early 1960s, economic growth has gathered pace, and industry started voicing demands in political arenas for more well-trained technicians. The post-war education system, which started with a simplified uniform model, was altered and a new line of vocational high schools (Colleges of Technology) was added (See Figure 2.3).

2.4 Colleges of Technology

By the 1960s, it became clear that there needed to be more well-trained people for the rapidly growing manufacturing industry in Japan. The Colleges of Technology (Kōtō Senmon Gakkō, *Kōsen*) were founded against this background to respond to these demands. They provide five-year engineering education for those who graduated from a junior high school. The first 12 national colleges were inaugurated in 1962, followed by a further 31 colleges established by 1965.

When they were established in 1961, the Association of Junior Colleges demanded that these new Colleges of Technology be treated differently from them. As a result, the Colleges of Technology were created originally as a completely separate type of higher education institution to train and foster engineers without giving their students clear access to further education opportunities. However, high levels of success in acquiring jobs for their graduates cemented their popularity, attracting a stable number of applicants, and also resulting in a further increase in the number of colleges across the country. In 1976, the government, under the premiership of Kakuei Tanaka, decided to found two national universities of technology which could accommodate graduates from Colleges of Technology. Subsequently, in 1992, the Colleges of Technology began to offer their students a two-year Advanced Course so that the graduates could be granted the opportunity to enter universities to obtain further degrees. Because of their origin, Colleges of Technology are still treated separately from 'universities and junior colleges' in official government data.

Currently, there are 51 national colleges, administered by the Institute of National Colleges of Technology, on top of three publicly-funded colleges (run by prefectures) and three private colleges, spanning six main specialties (Mechanical Engineering; Electrical and Electronics Engineering; Information and Computer Engineering; Chemistry, Biochemistry and Materials Science; Architecture, Civil and Environmental Engineering; and Business Administration). Annually, approximately 10,000 students enter the colleges, of whom 16 per cent are female ([9]).

The number of female students exceeded 1,000 in 1981 (2.2 per cent), 2,000 in 1986 (4.1 per cent), and 3,000 in 1989 (7.2 per cent). By 1991, it reached 5,856 (10.9 per cent). As of 2013, female students (9,681) account for 16.6 per cent. In Chemical Engineering, Civil Engineering, and Business Administration in particular, the ratios of female students began increasing since the 1990s and are now more than 40, 25 and 70 per cent, respectively.

At graduation, 10 per cent of (male and female) students progress to the two-year Advanced Course and receive a Bachelor's degree. With this qualification, approximately one-third of the students pursue postgraduate study at various universities, including the two universities of technology mentioned above. Although Colleges of Technology have been often overlooked in the shadow of mainstream higher education institutions (i.e. universities and junior colleges), they have provided both men and women with an avenue for pursuing science and engineering careers for more than 50 years.

The next section looks at how international pressure was brought upon Japan, which resulted in the passage of milestone legislation, the EEOL.

2.5 Equal Employment Opportunity Law

The first UN conference on Women was held in Mexico City in 1975 and received international attention. It declared the years 1976–1985 as the UN Decade for Women, and established a Voluntary Fund for the Decade. Subsequent conferences in 1980 (in Copenhagen) and in 1985 (in Nairobi) served as catalysts for an international movement for women's empowerment and equality.

In Japan, a survey conducted between 1982 and 1984 primarily highlighted the difficulties of working mothers (e.g. childbearing and child care) without making specific references to inequalities between men and women ([10]). In the period between the 1960s and the 1990s, few efforts were made by feminist scholars to strengthen the voices of those calling for equality. For instance, the Women's Studies Association of Japan was founded in 1979.

The EEOL was enacted in 1986 as a response to international pressures deriving from the UN convention on the elimination of all forms of discrimination against women ([11]). However, this law was controversial, as the primary political agenda in the late 1980s was the decreasing fertility rate and a crisis of the family. This was criticised as the main cause of this crisis. From this point of view, gender equality from a feminist perspective was constantly perceived as a possible threat that could further dismantle the cornerstone of Japanese society, i.e. family and marriage.

Exemplified by the notion of 'good wives, wise mothers', the difference between feminist movements in Western societies and Japan is often described in terms of the latter focusing less on individual autonomy and more on the family system. Buckley (1997: 278–279) argues that '(i)n any East Asian culture you will find that women have a very tangible power within the household. This is often rejected by non-Asian feminists who argue that it is not real power[2], but [...] Japanese women look at the low status attributed to the domestic labour of housewives in North America and feel that this amounts to a denigration of a fundamental social role – whether it is performed by a man or a woman'.

The original 1986 law was unpopular among most employers. They feared that a mandate to treat men and women equally at work would decrease productivity. It was criticised by the trade unions for not being far-reaching enough. The law only imposed the 'obligation to make an effort', and imposed no potential penalties on employers who ignored it. One of the consequences was that many employers began applying a new personnel management system by creating two main types of career path for new entrants. One is called 'executive-track position' (*it sōgō shoku*) and the other 'general-duties position' (*ippan shoku*).

As a result, discriminatory treatment between the two types of career was not sex-based, and therefore lawful, although more women opted for the latter track. Some complained that the law was inherently flawed because women faced harsher working conditions than they had previously encountered in order to compete with men. Many women were faced with the very hard choice between a marriage/family and a career. Even if they were willing to forgo/delay marriage to focus on a career, there was no guarantee that they would be able to equally compete with men in the executive-track position.

Despite its flaws, this law is viewed as a milestone for career-minded, working women in Japan, as a new path was opened to them.

The year 1986 was a significant time for domestic politics as well. Takako Doi became the first female party leader (of the Japan Socialist Party) in the country. She demanded that the Diet sign the UN Convention on the Elimination of all Forms of Discrimination against Women in 1985. The popularity she enjoyed was such that the Japan Socialist Party gained the highest number of seats in the 1990 Lower House Election. However, she stepped down as a leader in 1991, to protest the government's involvement in the (first) Gulf War.

The late 1980s and the early 1990s were rather turbulent for Japan in various ways. A recession followed the collapse of the economic bubble, and the selection ratio (the ratio of job positions versus that of job applications) worsened in the labour market. Several corruption scandals (e.g. insider trading and politicians' involvement in it) came to light. The LDP lost control of the government for the first time in 1993 since its 38-year reign began in 1955. Table 2.2 summarises the actions and policy measures taken in the United Nations and Japan since 1975.

2.6 Women's access to science and engineering research through higher education

In Japan, there are 783 four-year universities, of which 78 are women's universities, mostly focusing on liberal arts education. In this cultural context, women's universities in Japan continued to educate women in a manner consistent with the notion of 'good wives, wise mothers' (*ryōsai kenbo*). This strategy satisfied the needs of schools, companies or corporations, and in turn increased the schools' reputations in the long-term[3]. Women's universities continued to produce distinguished scientists and prominent alumni in various fields. Some of the well-known names include the mathematician Shihoko Ishii (TWCU, who received a

Table 2.2 Actions and policies taken in the United Nations and Japan since 1975

		United Nations	*Japan*
1975		– Establishment of International Women's Year by the General Assembly (Goals: Equality, Development and Peace) – The World Conference of the International Women's Year (Mexico City) – Adoption of the World Plan of Action	– Establishment of the Headquarters for Planning and Promoting Policies Relating to Women – Convening of the Council on Women's Issues
United Nations Decade for Women (1976–85)	1977		– Formulation of the National Plan of Action – Establishment of the National Women's Education Centre
	1979	– The 34th Session of the General Assembly of the United Nations – Adoption of the Convention on the Elimination of All Forms of Discrimination against Women	
	1980	– World Conference of the United Nations Decade for Women (Goals: Equality, Development and Peace) (Copenhagen) – Adoption of the Program of Action for the second half of the United Nations Decade for Women	
	1981		– Formulation of the priority targets for the second half of the period covered by the National Plan of Action
	1984	– ESCAP Regional Preparatory Meeting for the 1985 World Conference (Tokyo)	
	1985	– World Conference to Review and Appraise the Achievements of the UN Decade for Women (Goals: Equality, Development and Peace) (Nairobi) – Adoption of the Nairobi Forward-looking Strategies for the Advancement of Women	– Amendment of the Nationality Law (employing a bilinear system) – Promulgation of Act on Securing, etc. of Equal Opportunity and Treatment between Men and Women in Employment (hereinafter referred to as the Equal Employment Opportunity Law) – Ratification of the Convention on the Elimination of All Forms of Discrimination against Women

(*Continued*)

Table 2.2 Actions and policies taken in the United Nations and Japan since 1975 (Continued)

	United Nations	*Japan*
1986		– Expansion of the Headquarters for the Planning and Promoting of Policies Relating to Women (the organisation was expanded to all ministries and agencies; mandate was also expanded) – Convening of the Advisory Council to the President of the Headquarters for the Planning and Promoting of Policies Relating to Women
1987		– Formulation of the New National Plan of Action towards the Year 2000
1988	– CEDAW (The United Nations Committee on the Elimination of Discriminations against Women) 7th Session: Consideration of Japan's first periodic report on implementation of the Convention (New York)	
1990	– The 34th Session of the UN Commission on the Status of Women – Adoption of the First Review and Appraisal of the Implementation of the Nairobi Forward-looking Strategies for the Advancement of Women by the UN Economic and Social Council (ECOSOC)	
1991		– Promulgation of the Act on the Welfare of Workers Who Take Care of Children or Other Family Members including Child Care and Family Care Leave (hereafter referred to as Childcare and Family Care Leave Law)
1994	– ESCAP Regional Preparatory Meeting for the Fourth World Conference on Women (Jakarta) – The International Conference on Population and Development (ICPD) (Cairo) – Adoption of the Cairo Declaration and Programme of Action	– Establishment of the Office for Gender Equality, the Council for Gender Equality and the Headquarters for the Promotion of Gender Equality

	– CEDAW 13th Session: Consideration of Japan's second and third periodic reports on implementation of the Convention (New York)	
1995	– The Fourth World Conference on Women (Beijing) – Adoption of the Beijing Declaration and Platform for Action	– Revision of the Child Care and Family Care Leave Law (legal establishment of the Family Care Leave System)
1996		– Establishment of the Liaison Conference for the Promotion of Gender Equality (Egalite Network) – Formulation of the Plan for Gender Equality 2000 – The National Plan of Action for Promoting a Gender-Equal Society by the Year 2000
1997		– Establishment of the Council for Gender Equality by Law – Revision of the Equal Employment Opportunity Law and related laws – Promulgation of the Long-Term Care Insurance Act
1999	– ESCAP High-level Intergovernmental Meeting (Asia-Pacific Regional Preparatory Meeting for Women 2000 General Assembly) (Bangkok)	– Promulgation and enforcement of the Basic Law for a Gender-Equal Society – Promulgation and enforcement of the Food, Agriculture and Rural Areas Basic Act
2000	– The 23rd Special Session of the United Nations General Assembly 'Women 2000' (New York)	– Formulation of the Basic Plan for Gender Equality
2001		– Establishment of the newly formed Council for Gender Equality and the Gender Equality Bureau – Enforcement of the Act on the Prevention of Spousal Violence and the Protection of Victims – The first Gender Equality Week – Cabinet Decision on the Policy on Support Measures for the Balancing of Work and Child Raising – Campaign for Eliminating Violence against Women decided by the Headquarters for the Promotion of Gender Equality

(*Source*: [21])

doctoral degree in science from Tokyo Metropolitan University), earth scientist Katsuko Saruhashi (Imperial Women's College of Science, who received a doctoral degree in science from the University of Tokyo), and chemist Reiko Kuroda (alumna of Ochanomizu, who received a doctoral degree in science from the University of Tokyo). Graduates from those women's colleges normally pursued their graduate studies either at one of the renowned national universities or universities outside Japan. Therefore, in the realm of scientific research, the paths that were taken by their predecessors were followed for a long time.

2.7 Underrepresentation of women in STEM

The issue of the underrepresentation of women in STEM was already recognised by the Science Council of Japan (advisory body to the Prime Minister, see the section below for more details) in the mid-1970s. Yet there was no debate about how male dominance in STEM subjects could practically be altered in the education and industry sector. The 1986 Law for an Equal Employment Opportunity for Men and Women therefore opened this debate by officially recognising women as a legitimate constituent of the workforce for the first time. Figure 2.4 shows the number of newspaper articles which mentioned 'women scientists' (Jyosei Kagakusha), covering the period of 1980 and 2012. This includes all the major four newspapers (*Yomiuri, Asahi, Nikkei* and *Mainichi*), and Nikkei's specialist daily newspaper featuring industry (*Nikkei Sangyō Shinbun*). The total number of articles tripled in 1988, and increased further into the 1990s.

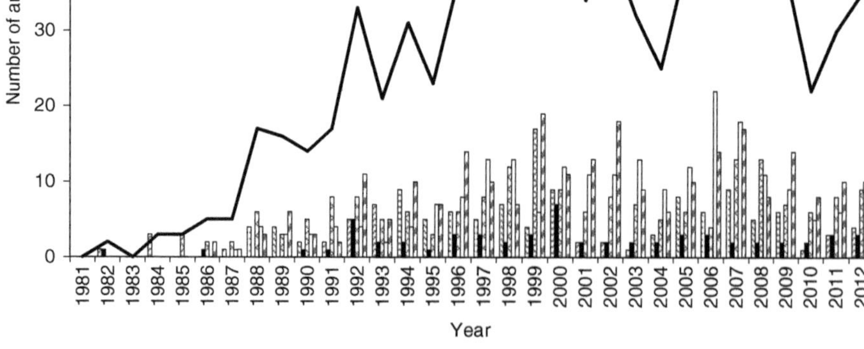

Figure 2.4 Number of newspaper articles mentioning 'female scientists', 1981–2012.

Source: Yomiuri, Asahi, Mainichi, Nikkei and Nikkei Sangyō Shinbun.

It would be safe to assume that the 1986 EEOL was a critical juncture, which heightened the awareness of the public towards women scientists. However, Japan did not see a series of government initiatives in women's education until the 2000s. As a result, a two-decade gap exists between Japan and most other European countries in rectifying gender disparities in society, including in STEM education. By the 2000s, the pool of many successful female scientists was concentrated in women's universities in the case of Japan. Women's studies institutions such as the Centre for Women's Studies (founded in 1990) at TWCU and Tsuda College began to feature some prominent alumnae in STEM in their publications. *Women's Self-support and Science Education: A History of the Science Department of Tsudajuku College* was published in 1987. At Ochanomizu University, the Institute for Gender Studies was established in 1996, and published the *Catalogue of Chika Kuroda's Archives* in 2000. At this stage, efforts were still primarily focused on advertising the past achievements of great female scientists that each women's university produced to differentiate themselves from other types of universities and attract more students.

Yet some more structural changes began to be observed in response to the changing social mood. As previously mentioned, more girls started to choose Colleges of Technology around that time. In 1992, the Faculty of Home Economics at JWU was finally reorganised into the Faculty of Science, following the two national women's universities (Ochanomizu and Nara, both with the Faculty of Science established in the 1950s).

In 1995, the enrolment rate at junior colleges and universities almost equalised at about 20 per cent, and that of the former began decreasing after this time. In 1996, the number of women who went to a four-year university (24.6 per cent) exceeded the number of those who went to a two-year college for the first time.

Since the 1960s, the demand for university education in Japan has increased greatly. In recent years, more than 50 per cent of both men and women enrol in universities. For women, a great shift from junior colleges to four-year universities has occurred. In previous situations where women's exit from the labour market for marriage, childbirth and childcare was normal (shown in the well-known M-shaped curve in the labour force participation rate curve), university education was regarded as a consumption good rather than an investment good.

However, changes in the employment environment for women after the EEOL brought about a shift from a junior college model (consumption goods) to a university model, i.e. four-year degree course (investment goods) ([14]: 108). Higher education that one purchases started to be seen as a good that pays dividends and provides great benefits for many years into the future, rather than a good that offers instant gratification and utility.

At the turn of the century, the MEXT reviewed the progress in the 2002 Annual Report on Science and Technology, and stated that 'the number of female researchers increased rapidly after the EEOL was enacted in 1986, but the increase became stagnant in recent years'. ([15])

The next section looks at the communities of scientists, and how women scientists have been represented since WWII.

2.8 The Science Council of Japan

The Science Council of Japan (SCJ) was established in January 1949 as a special advisory body to the Prime Minister. The idea was recommended by the GHQ, and the Diet passed the relevant legislation in July 1948. The mission of the SCJ is to enhance and promote the field of science, and it operates independently of the government. It has 210 council members, and another 2000 affiliate members are selected from academic communities ranging from humanities, social and natural sciences to engineering.

The SCJ was originally divided into seven sections (Literature, Law, Economics, Science, Engineering, Agriculture and Medicine) reflecting the same divisions in former Imperial Universities. Members were democratically elected by scientists, and the original function of the SCJ was to provide advice on science policies to the government. One term normally runs between three and four years.

Over time, the SCJ lost influence over policy-making, including in terms of budgetary allocation, particularly after the Ministry of Education established its own Science Council in 1967. In 1983, the government amended the law, changing the method of selection of members from direct election to the appointment by the Prime Minister (based on shortlists provided by major academic societies). However, the SCJ was considered to be the highest policy body of the scientific community at the national level until the 2001 reorganisation of government ministries led to the creation of the Council for Science and Technology Policy. At the same time, the roles of and selection processes for the SCJ were reviewed and then reorganised in 2005. Since then, the SCJ has had three sections (Section 1: Humanities and Social Sciences, Section 2: Life Sciences, and Section 3: Physical Sciences and Engineering), replacing the former seven divisions.

The history of female members of the SCJ is worth noting. The first female member, who was elected for the 12th SCJ in 1981, was an earth scientist, Katsuko Saruhashi (only one woman was elected). She was a member for one term (1981–1985). In the following term (13th· 1985–1988), three female scholars were elected, two of whom were social scientists (Yasuko Ichibangase and Etsuko Yasukawa). Masako Hayashi was the only natural scientist (architecture, JWU graduate). The number of women in the SCJ did not increase in the following terms, as Table 2.3 shows.

The number of members selected from the natural sciences stayed at one (0.5 per cent) from 1981 until 1997. Even the total number remained negligible, fluctuating between one and three (1.5 per cent), until the SCJ changed its selection rules in 2005. The new rules were aimed at increasing the proportion of both female and younger members selected from academic societies. As a consequence,

Table 2.3 Female members of the Science Council of Japan (by number and discipline)

	Female members (STEM only)	*No. of female members*	*% of female members*
12th (1981–1985)	Katsuko Saruhashi (earth scientist)	1	0.5%
13th (1985–1988)	Masako Hayashi (architect)	3	1.4%
14th (1988–1991)	Masako Hayashi (architect)	2-3	1.0–1.4%
15th (1991–1994)	Masako Hayashi (architect)	2-3	1.0–1.4%
16th (1994–1997)	Junko Shimada (dietician)	1	0.5%
17th (1997–2000)	Junko Shimada (dietician)	2	1.0%
18th (2000–2003)	Fumiko Yonezawa (physicist) Masako Niwa (life scientist)	6	2.9%
19th (2003–2005)	Kimiko Murofushi (biologist) Mizuho Ishida (earth scientist) Tomoko Ogawa (biologist) Ikuko Ezawa (dietician)	13	6.2%
Reformed SCJ	*Number only in STEM fields*		
20th (2005–2008)	10 Life Sciences 7 Physical Sciences & Engineering	42	20.0%
21st (2008–2011)	11–12 Life Sciences 8–9 Physical Sciences & Engineering	43-47	20.5–22.4%
22nd (2011–2014)	13 Life Sciences 10 Physical Sciences & Engineering	49	23.3%
23rd (2014–present)	16 Life Sciences 10 Physical Sciences & Engineering	49	23.3%

Note: Substitute members are included in the total number, which explains the range.
(*Source*: Data provided by the SCJ. Adapted by the authors.)

the number of members selected from natural sciences quadrupled in 2005, from four to 17. The total number of female members also tripled (from 13 to 42).

A member can serve for a maximum of three terms (12 years). Since the first female member was elected in 1981 up to September 2014, there have been 41 female members in STEM fields. Out of those, nine members have their first degrees from the University of Tokyo and Ochanomizu University, while three are graduates from Kyoto University. Two female members came from each of the following: Nara Women's University, JWU, Hokkaidō University and Kyūshū University. In terms of doctoral degrees, more than half (22 members) obtained their degrees from the University of Tokyo, followed by four from Kyoto University, and two each from Hokkaidō, Kyūshū and Nagoya. Seven out of nine graduates from Ochanomizu went to the University of Tokyo. The path that their predecessors took since the early twentieth century has continued.

However, the reform of the selection process did not progress naturally. In 1994, the sixteenth term of the SCJ had only one female member in total, i.e. it actually decreased from the two of the previous term. It was a critical moment for female scholars from all sciences, as the SCJ (in its fifteenth term) had just made an urgent call for improving the conditions for female scientists.

In response to this, a new group called the Japanese Association for the Improvement of Conditions of Women Scientists (JAICOWS) was formed, with former member of the SCJ and sociologist, Ichibangase, becoming its first president. JAICOWS, as a voluntary non-profit group loosely affiliated to the SCJ, began campaigning for its cause, particularly demanding an approach in line with the government's policy of equal participation between men and women. Through various symposia and deliberations, in 2000, the SCJ set the goal of increasing the ratio of female members up to ten per cent by 2010.

2.9 Conclusion

This chapter has described how post-WWII Japan revamped its education system, widening the participation of women in tertiary education as equals with men, and how it promoted co-education. However, the curriculum was still designed differently for boys and girls at the secondary level, and parents also believed that sending their daughters to a co-educational, former Imperial University remained socially risky, as it can jeopardise their future marriage and family life.

Although the underrepresentation of women in STEM subjects was well known by the 1970s, the situation did not improve until three conditions were met. First, international pressure had to be exerted on the government to deal with gender inequality. Second, there had to be social movements and awareness campaigns around gender inequality beyond academic research. Finally, strong initiatives were required of the academic community.

It would require another two decades before these three 'streams' became joined, and the government policy was finally changed in the early 2000s. Before then, the declining economic power and fertility rate were yet to be discovered, and traditional policy-making styles questioned in the 1990s. Although the underrepresentation of women in STEM will remain unresolved, many struggles shared by other working women will be highlighted in the next ten years following the EEOL. By the end of the 1990s, the Basic Act for Gender Equal Society (to give its official English title, but in Japanese, 'Basic Act for Promoting Equal Participation of Men and Women' (*Danjyo Kyōdō Sankaku Kihonhō*)) was to be enacted. The next chapter looks at policy developments towards the enactment of this Basic Act, as well as the impact of the EEOL on women's education and labour market.

Notes

1 As of 2009, female students at undergraduate level account for 24.4 per cent at Tōhoku University, 29.6 per cent at Kyūshū University, 27.3 per cent at Hokkaidō University, 31.7 per cent at Nagoya University and 35.7 per cent at Osaka University.

2 Some scholars refer to this situation as 'domestic matriarchy' ([12]).

3 A good indicator of the reputation of traditional women's universities is the record of overall success for their graduates in the job market. The ratio of graduates who started off their careers in large corporations such as Hitachi, Tōshiba, Fujitsū and Canon is known to be very high (roughly thirty per cent). In 2007, among the top twenty in this league table, four were women's universities ([13]).

References

[1] National Institute of Science and Technology Policy, MEXT. Education reform for the 21st century: An introduction postwar education reform in retrospect. [online]. Available from: http://www.mext.go.jp/b_menu/hakusho/html/hpac200101/hpac200101_2_004.html, 2001. Accessed 20 September 2014.

[2] A. Tsugawa and Y. Kanomi. *Hiraku: Nihon no josei kagakusha no kiseki (Opening: Tracing the paths of Japanese female scientists)*. [In Japanese] Domesu Publishing, 1996.

[3] T. Yukawa. Sengo no kyūtaiseika ni okeru jyoshi daigaku setsuritsu kōsō ni kansuru ichikōsatsu: kyōiku rinen to gakubu kōsei wo chūshin ni (analysis of the initiative of establishing women's universities under postwar former school system: focusing on their education philosophy and constitutive faculties) [In Japanese]. *Waseda University Academic Research (Humanities and Social Sciences) Department Bulletin*, (62): 35–51, 2014. [online] Available from: https://dspace.wul.waseda.ac.jp/dspace/bitstream/2065/41358/1/GakujutsuKenkyu_Jinbun_62_Yukawa.pdf. Accessed 24 November 2014.

[4] T. Kobayashi. *Tōdai gōkaku kōkō seisui-shi (Ebb and flow of high schools' reputation: record of success in sending students to the University of Tokyo)*. [In Japanese] Tokyo: Kōbunsha Shinsho, 2009.

[5] K. Yawata. *47 Todōfuken no Meimon Kōkō (Prestigious high schools in 47 prefectures)*. [In Japanese] Tokyo: CDI Publishing, 2008.

[6] T. Hirano and K. Hirai. Jyosei wo meguru shakaiteki kankyō no rekishiteki tenkai: Jyoseishi nenpyō no kisai kōmoku kara (Historical development of women's social environment: Extract from a chronology of women). Technical Report 23, Teikyō Shakaigaku, 2010. [online]. Available from: https://appsv.main.teikyo-u.ac.jp/tosho/thirano_khirai23.pdf. Accessed 24 November 2014.

[7] Y. Ichibangase and E. Okuyama. *Fujin kaihō to jyoshi kyōiku (Women's liberation and education)*. [In Japanese] Tokyo: Keisō Shobō, 1975.

[8] T. Kamiko. *Chūgakkō kyōkasho no genjyō: gijyutsu/katei* (current situation in junior high school textbooks: technical arts and home economics), chapter in *Gender Discrimination in Text Books* (Kyōkasho no naka no danjyo sabetsu) [In Japanese], pages 175–184. Tokyo: Akashi Shoten, Yoshinori Ito, Masako Owaki, Tatsuko Kamiko, and Mutsuko Yoshioka edition, 1991.

[9] National Institute of Technology. *Kōsen Joshi Hyakka Jr.* (The Data Book on Girls of National Colleges of Technology) [In Japanese]. Nara: National Institute of Technology, 2012.

[10] M. Kuwahara. Japanese women in science and technology. *Minerva*, 39(2): 203–216, 2001.

[11] B. Mello and M. Strausz. International norms and women's rights in Turkey and Japan. *Journal of Women, Politics & Policy*, 32(4): 333–362, 2011.

[12] S. Iwao. *The Japanese Woman: Traditional Image and Changing Reality*. New York: Free Press, 1993.

[13] The Weekly Economist. Daigaku shūshoku-ritsu rankingu (League table of universities: the top 150 based on their graduates' performance of entering successful large corporations) [In Japanese]. In *The Weekly Economist*. Tokyo: Mainichi Shinbun Publishing, 14/21 August 2007.

[14] Y. Tanaka and Y. Chiba. Jyoshi no daigaku kyōiku jyuyō to koyō kankyō no henka: 4 nen sei to tanki daigaku no shiganritsu kettei no nikō/takō rōjitto moderu ni yoru jikeiretsu

bunseki (Time series analysis of binary-multinomial logit model of volunteer rate determination of junior college and 4 year university in changing job market and demands for tertiary education for girls) [In Japanese]. *Kyoto Sangyō University Journal*, pages 89–113, 2011. [online]. Available from: http://ksurep.kyoto-su.ac.jp/dspace/bitstream/10965/460/1/AHSUSK_SSS_28_89.pdf. Accessed 15 November 2014.

[15] National Institute of Science and Technology Policy, MEXT. Annual report on the promotion of science and technology 2002 [In Japanese]. Technical report, MEXT, 2002. [online]. Available from: http://www.mext.go.jp/b_menu/hakusho/html/hpag200201/index.html Accessed 13 December 2014.

[16] Office for Gender Equality, the University of Tokyo. *Jyoshi Tōdaisei no ayumi* (History of Female Students at the University of Tokyo) [In Japanese] [online]. Available from: http://kyodo-sankaku.u-tokyo.ac.jp/activities/model-program/library/UTW_History/Page03.html. Accessed 24 November 2014.

[17] Kyoto University Gender Equality Promotion Center. *Suuji de miru jyosei kenkyūsha* (Female Researchers: the Figures) [In Japanese] [online] Available from: http://www.cwr.kyoto-u.ac.jp/tokei/. Accessed 24 November 2014.

[18] Ministry of Education, Culture, Sports, Science and Technology. *Gakkō Keitōzu* (Diagram of school system) [In Japanese]. [online]. Available from: http://www.mext.go.jp/b_menu/hakusho/html/others/detail/1318188.htm. Accessed 14 December 2014.

[19] T. Tachibanaki. *Jyosei to gakureki: Jyoshi kōtō kyōiku no ayumi to yukue* (Women and educational background: Historical development of higher education for women and its future) [In Japanese] Tokyo: Keisō Shobō, 2011.

[20] Ministry of Education, Culture, Sports, Science and Technology. *Gakkō kihon chōsa* (School basic survey) [In Japanese]. [online]. Available from: http://www.mext.go.jp/b_menu/toukei/chousa01/kihon/kekka/1268046.htm, 2013. Accessed 17 November 2014.

[21] Cabinet Office. *Actions Taken in Japan and the United Nations since International Women's Year* [online]. http://www.gender.go.jp/english_contents/pr_act/pub/pamphlet/women-and-men12/pdf/3.pdf, 2012. Accessed 21 April 2015.

3 From equal opportunities to equal participation

Women's careers in science and engineering in the 1990s

3.1 Introduction: winds of change from the mid-1980s onwards

The Equal Employment Opportunity Law (EEOL) was highly controversial in Japan, as it was a rather reactive move in response to international pressure, and Japanese society was not prepared to embrace the implications. Some proponents even argued that the EEOL's positive impact on women's employment was too little, which underlines the fact that many companies simply created a two-track career system to block women's entry to a professional career on equal footing with men. Nevertheless, there was no doubt that the timing of the enactment coincided with the end of many social customs under the post-war regime in Japan.

With the EEOL, women were officially and legally allowed to work on an equal footing with men. Generally, conservative attitudes of both men and women in Japanese society hardly changed, but women slowly began to proceed to higher education institutions of various kinds, including Colleges of Technology. In an ever more competitive global market, some firms began to value talent, irrespective of gender, treating men and women equally ([1]).

After a mild economic downturn in the mid-1980s, Japan's economy picked up again and the rate of economic growth was on average 5 per cent up to the early 1990s, when a long recession started. In the 1990s, the bubble economy burst and the LDP single-party government collapsed. Scepticism towards the Japanese-style high economic growth model emerged. Against this background, the 'lost decade' began, and the concept of 'gender equality' was placed firmly on the agenda again by the 4th World Conference on Women in 1995. While the word 'gender-free society' caused stirs and controversies, gender equality infiltrated steadily.

This chapter summarises the changing political and social landscapes in the 1990s when a number of policy initiatives were taken. It outlines traditional Japanese policy-making styles and what changes took place around those structures, which later contributed to policy changes, including the enactment of the BAGES in 1999. This chapter also looks at how the 1986 EEOL became a critical juncture, resulting in changes in the choice of educational institutions and professional careers for women in science and engineering.

3.2 Changing political and social landscapes in the 1990s

Under the so-called 1955 regime, the LDP governed Japan for 38 years without a change of government until 1993, while the Japan Socialist Party was in permanent opposition. The days of LDP one-party rule were characterised by the 'iron triangle' of law-makers, bureaucrats and industries ([2]; [3]). Factions (*habatsu* in Japanese) played a central role in keeping the LDP in power for 38 years. Factions were formed by like-minded people, not necessarily distinguished by policy differences, as the main functions were power-brokering. Therefore, a faction would be typically headed by powerful politicians who were considered potential party-leader candidates. The factions relied on a complex network of patron–client (*oyabun–kobun*) relationships, and rival factions in power struggles sought to serve their groups' interests first, rather than the party's. This often created a mutually beneficial pact in which clients (junior members) provided full support for their patron's future bid for party leadership (i.e. premiership). In return, the patron offered his clients funding for electoral campaigns and backing. Faction-based politics in the LDP era successfully brought about a repeated pseudo-change of government ([4]: 18). However flawed as a democratic polity, the system proved efficient in keeping power contests out of policy disputes ([5]: 140). Yet, in the early 1990s, the negative side became apparent through a series of money scandals in electoral campaigns ([6]). This system, underpinned by the iron triangle, sustained an old boys' network, not leaving much room for women to enter.

Bureaucrats similarly nurtured a culture of serving their own bureaus' interests first. The culture was cynically ridiculed by the saying 'bureau's interests before ministry's, section's interests before bureau's'. This became widely known as '*Tatewari Gyōsei*' (administrative system rigidly segregated along the lines of each ministry). Politicians sharing an interest in a particular kind of policy or sector became known as 'tribe' (*zoku*) politicians, who built close, long-term alliances with bureaucrats. Becoming a *zoku* politician was considered to be an important step in one's career, with such politicians acting as coordinators in various political-economic transactions ([4]).

They defended the interests of bureaucrats as well as interest groups affected by the particular ministry. Examples of 'welfare expert' (Kōsei Zoku) politicians includes Ryūtarō Hashimoto (Minister for Welfare, 1978–1979; Prime Minister, 1996–1998) and Jun-ichirō Koizumi (Minister for Welfare, 1988–1989, 1996–1998; Prime Minister, 2001–2006). In Education, Toshiki Kaifu (Minister of Education, 1976–1977, 1985–1986; Prime Minister, 1989–1991) and Yoshirō Mori (Minister of Education, 1983–1984; Prime Minister, 2000–2001) are well-known.

The iron triangle had both positive and negative consequences. On the positive side, the government was stable throughout the period between 1955 and 1993, during which a myth was created to the effect that a change of government has a negative impact on economic growth. It was tenaciously held on to and the maintenance of a long and stable LDP government became a common goal and a supreme priority of the overall electorate. A coalition government was considered to be something to be avoided, as it would bring about a vicious cycle of 'coalition (short-lived government) a hindrance to economic prosperity' ([4]: 18–19).

However, the cracks in the old system began to emerge in the early 1990s, when the trading and corruption scandals (notably, surrounding a human resources company called Recruit) were exposed in the late 1980s and the bubble economy burst in 1991. In the 1993 general election, the LDP lost its overall majority and was not able to form a government for the first time since 1955. Subsequently, an eight-party coalition government was formed, but collapsed after 10 months, passing major legislation regarding the electoral system reform. Japan's first minority government ensued in April 1994, lasting only two months. The Japan Socialist Party (now the Social Democratic Party of Japan[1]) Tomiichi Murayama then became Prime Minister, leading another coalition government with the LDP and a small party called *Sakigake* (Harbinger Party). This government and the following coalition governments created an environment in which the concept of 'equal participation of men and women' (*Danjyo Kyōdō Sankaku*, translated into English simply as Gender Equality) became a law in 1999.

3.3 Towards the Basic Act for Gender Equal Society

In 1990, the first review and appraisal were conducted regarding the implementation of the Nairobi Forward-looking Strategies for the Advancement of Women by the UN Economic and Social Council (ECOSOC). In response, the government set up an expert group which put forward a series of recommendations to improve the situation surrounding women in Japan. The New National Action Plan towards the Year 2000 (First Revision) was agreed by the Cabinet in May 1991. 'Equal participation of men and women' (*Danjyo Kyōdō Sankaku*) appeared for the first time in the official document of the central government. In December 1992, a new ministerial portfolio (Minister for Women's Affairs) was created and added to the role of Yōhei Kōno, then Chief Cabinet Secretary.

In August 1991, the Council for Deliberation on Planning and Promoting Policies relating to Women's Affairs was convened under the Cabinet Office (then called the Prime Minister's Office). The then Councillor of Cabinet Secretariat, Kanji Takaoka was informally approached and asked by one of the Council members, Makiko Arima as to whether a basic act could be legislated ([7]: 6). Arima was a journalist and representative of the Japanese government to the UN Commission of the Status of Women (and an alumna of Tsuda College).

Against this backdrop, the new coalition government, formed under Murayama in 1994, upgraded the Headquarters for Planning and Promoting Policies Relating to Women's Affairs (established in 1975) to the Office of Gender Equality (to give its official English title), and placed it within the Cabinet Office in 1994. The Japanese title for this office was *Danjyo Kyōdō Sankakushitsu* (Office for

Promoting Equal Participation of Men and Women). The Council for Deliberation on Planning and Promoting Policies relating to Women's Affairs also became the Prime Minister's Council for Gender Equality (*Danjyo Kyōdō Sankaku Shingikai*) by Cabinet Ordinance. The first chairperson was Yōko Nuita, a journalist and alumna of Tsuda College, and the first female senior commentator on NHK (*Nihon Hōsō Kyōkai*, Japan Broadcasting Corporation).

In the same year, consulted by Prime Minister Murayama, Mariko Arima and another member of the Prime Minister's Council for Gender Equality, Genrokurō Furuhashi (former Permanent Vice-Minister, Management and Coordination Agency) embarked on discussions about the possibility of legislating a basic act. ([7]: 7)

Following The Beijing Platform for Action signed by 189 governments at the 4th World Conference on Women in 1995, the Prime Minister's Council for Gender Equality published a policy consultation report entitled 'Vision for Equal Participation of Men and Women' (*Danjyo Kyōdō Sankaku Bijon*) in July 1996.

It was the first time the word 'gender' (*jendā*) was used in an official government document. The document states that 'this report seeks to achieve a society where women and men can take part and make individual contributions based on their personalities, not bound by their genders formed socially or culturally'. ([7]: 10) This report laid the foundation for the 1999 BAGES.

Although the second Hashimoto LDP-led minority government (November 1996–July 1998) was committed to the cause, the concept of 'equal participation of men and women' is much more nuanced than 'gender equality'. The commitment was also a by-product of coalition politics. Since the stability of the LDP government relied on support from the Socialist Party and Sakigake (Harbinger), the tripartite agreement included the promotion of this policy ([8]). A member of the House of Councillors (*Sakigake*), Akiko Dōmoto[2] is one of those strong proponents behind this agreement. ([7]: 36)

The ambivalence among conservative party members towards the idea was clear in the parliamentary debates that preceded the passage of the new law. The opposition party member Yuiko Matsumoto (Democratic Party of Japan) raised a question in the House of Representatives Cabinet Committee on 6 March 1996, highlighting a difference between the government's final plan (The 2000 Plan for Equal Participation of Men and Women) and some policy consultation documents (notably Vision for Equal Participation of Men and Women) in the way in which 'gender' was conceptualised. Notably, the terms 'gender free society' or 'gender equality' were not included as ultimate goals. The tenacity of the 'family' perspective led to a strong resistance to any commitment to gender issues in a wider political context.

However, the Prime Minister's Council for Gender Equality was re-established, this time by law (Act for Establishment of Gender Equality Council), which was passed in March 1997. The concept of Equal Participation of Men and Women was defined in a manner similar to the 'Vision for Equal Participation of Men and Women'. Professor Sumiko Iwao[3] was elected as the first Chair. In November 1998, the Council submitted a policy consultation report, which essentially became the Bill of the Basic Act. The Bill was agreed by the Cabinet in February 1999, and submitted to the Diet. It was passed by the House of Councillors in May and by the House of Representatives in June, unanimously by both. The word 'gender' (*jendā*) was not included in the law.

The Basic Act for Gender Equal Society (*Danjyo Kyōdō Sankaku Kihonhō*) was enacted in June 1999 (see Appendix A.1). The Law had five principles: (i) the establishment of human rights, (ii) further democratisation through active participation in policy making, (iii) the promotion of a deeper understanding and acceptance of gender-sensitive perspectives, (iv) the creation of new value system, and (v) to make contributions to a global society. Articles 8 and 9 of the BAGES stipulate that the central and local governments are responsible for the comprehensive formulation and implementation of policies related to promoting the formation of a gender-equal society.

In 1999, the Cabinet Office launched annual surveys on the perceptions of the general public about the issue of gender equality ([7]). In June 2000, the 23rd Special Session of the United Nations General Assembly, 'Women 2000', was held in New York. In December 2000, the Basic Plan for Gender Equality was formulated, following the BAGES. The Basic Plan for Gender Equality outlined concrete measures to support and realise the BAGES (see Appendix A.2).

Since the ground-breaking legislation was passed, 'equal participation between men and women' has become a widely used phrase nationwide, as local governments and municipalities had the obligation to formulate by-laws. However, this did not mean that the concept was widely accepted and supported. A backlash started, with many people (both men and women) expressing apprehension and opposition to the idea. ([7], 33)

Just as the 1986 EEOL was met with divided opinions, the road to equal participation of men and women in the labour market was long and arduous. A number of policy initiatives were required. The next section examines some of the new policy initiatives to support working women and encourage childbearing.

3.4 Japanese welfare state: familialism and the male-breadwinner model

The Japanese welfare state is characterised as a hybrid of Liberal and Conservative regimes, with a strong flavour of familialism. Familialism is 'an ideology that promotes family as a way of life and as a force for social integration' ([9]: 142). In Japan, taxes have been kept low and, hence, wages and employment levels have been maximised, while there is universal healthcare coverage based on the social insurance model. In the labour market, there has been a strong gender role division, and labour force participation rates among women have been among the lowest in advanced industrial economies.

In particular, Japan is known as a male-breadwinner model. The concept of the breadwinner model originates from patriarchical norms. One member of a family earns the money to support the others. Usually, the male working outside the home provides the family with income and work-related benefits, while the female stays at home as a housewife, providing care for their children and the elderly (hence, the name, male-breadwinner model).

Apart from the presence of *familialism*, the characteristics that are typically associated with the male bread-winner model include limited development of family policies, permanence of family solidarity and dependence, limited flexibility of the labour market and unequal division of family work between men and women. As a result, there is limited participation of women in the labour market ([10]).

The Japanese welfare state is often classified as an East Asian or Developmental regime ([11]; [12]; [13]). Social policy has long been regarded as secondary to the country's primary goal, namely, economic growth. However, social investment is seen as a key productive factor for economic growth, the enhancement of political stability and governmental legitimacy.

It is therefore interesting to see that in Japan the nature of social policy began to change in the early 1990s, when the traditional economic model ceased to work, in addition to the emergence of the paradoxical combination of strong familialism and very low fertility rate ([14]; [15]).

3.5 Support for working mothers

The Japanese government established child allowances in 1972. Originally only large families were entitled to the allowances, which covered the third and subsequent children below 18 years of age. The entitlement was extended to include the second child in 1988, and subsequently all children in 1994 (under the age of three). The age restriction was relaxed in December 1999, as a result of a coalition agreement between the LDP and Kōmei Party (affiliated with the Sōka Gakkai Buddhist sect) ([16], 132; [17], S69).

Under the Labour Standards Act, women are entitled to a fourteen-week maternity leave, six weeks before the baby is born and eight afterwards. They are entitled to 60 per cent pay during this leave, and another 25 per cent as an additional maternity allowance (this took effect in April 1999). On top of this, a lump sum payment is added that generally covers the direct costs of a normal birth ([16]: 140).

In Japan, the Child Care and Family Care Leave Act was passed in 1992. The law provided a year-long job-protected leave for parents taking care of infants and mandated that they should be paid 25 per cent of their normal pay. In April 2000, the amount was increased to 40 per cent of usual pay. However, the law initially did not work in practice, with very low uptake rates. 56.4 per cent of female employees who were eligible to take leave did so, and only 0.42 per cent of eligible male employees. As with the EEOL, it was based on 'Administrative Guidance', without meaningful enforcement mechanisms including lawsuits. The leave-takers were regarded by their boss and colleagues as 'slackers', and they worried about the security of position after their return ([16]: 140).

A raft of policy initiatives were introduced during this period in response to the '1.57 shock' total fertility rate drop in 1989. These included the 'Angel Plan' (1994) and the 'New Angel Plan' (1999), both of which were designed to make having children an easier and more attractive option. Collaboration among multiple ministries, local and municipal governments was sought for the implementation of these policies. Additionally, an emergency supplementary budget was allocated to build more childcare centres. Furthermore, the long-term care insurance system was established in 2000. These policies were aimed at shifting responsibility away from the private sphere (shouldered primarily by women in the household) into the public domain.

3.6 Revision of the Equal Employment Opportunity Law

As previously noted, opinions have been divided concerning the impact of the EEOL enacted in 1986, though there is no question that it was a milestone for aspiring professional women. Some argue that the EEOL had positive impacts on working women, while others emphasise the defects of the law ([18]). On the positive side, since the early 1990s, firms such as Shiseidō (cosmetics), IBM Japan, and Asahi Kasei (chemical products) began openly treating female employees in a fair manner, encouraging them to take maternity leave. They also introduced policies to hire more women ([7]: 80–83).

When the EEOL was enacted, the Labour Standards Law, originally enacted in 1947, was also amended. The 1947 law restricted women's overtime working hours to two hours per day, six hours per week, and 150 hours per year. Holiday work and late-night work (from 10 p.m. to 5 a.m.) were also prohibited. In 1986, these restrictions were abolished only for women in managerial positions and in 14 occupational categories requiring specialised knowledge or technical skills. However, since there were only a few women working under these conditions exempted from the restrictions, it is estimated that fewer than 5 per cent of all female employees worked on a totally equal basis with men, even after the enactment of the revised law ([19]: 105–106).

At the ten-year review of the EEOL in 1995, the government decided to amend the EEOL alongside the Labour Standards Act in 1997, and the above-mentioned Child Care and Family Care Leave Act was also revised (all three enacted in 1999). Guidelines were issued to provide employers with appropriate measures to take with regard to recruitment, hiring, assignment, promotion, training, firing and retirement. The main change was to give the law stronger enforcement mechanisms by switching from non-binding targets to the prohibition of discriminatory approaches based on gender.

The 1997 amendment of EEOL also recognised sexual harassment for the first time as a form of gender discrimination, and placed an affirmative duty on employers to prevent sexual harassment in the workplace. The new EEOL granted the employee the right to force her employer into mediation. The Equal Opportunity Mediation Commission was created, and the Ministry of Labour (now MHLW) was given authority to release to the media the names of the companies who failed to comply with the agency's advice, guidance, and recommendations regarding discriminatory practices. ([20]: 559)

However, in practice, the male breadwinner model was sustained through the income tax law. Significant tax deductions for a dependent spouse (*Haigūsha Tokubetsu Kōjo*, spousal deduction) are provided in Japan. These deductions are designed to support families by supplementing the earnings of the primary wage earner. When the secondary earner's income exceeded the ceiling (1,030,000 JPY), there was no deduction. Labour economists demonstrated that many women chose to work part-time, and accepted low-paid work for this reason. ([21] [22]: 170–171) Feminists also have long criticised this tax scheme as a great disincentive for women who might otherwise consider taking part in the labour market on equal terms with men ([16]: 134). The issue of abolishing this tax scheme has been raised several times since 2000, although it has not yet been achieved.

On the negative side of the EEOL, the two types of career path for new entrants: 'executive-track position' (*sōgō shoku*) and 'general-duties position' (*ippan shoku*) were created, as mentioned in Chapter 2. Many firms remained hesitant about hiring women, and women were faced with the very hard choice between marriage/family and a career. Even when they decided to focus on a career, in most cases, they were unable to compete with men on the executive-track position ([23]).

According to the 2011 study by the MHLW (male n=1313, female n=235), among the women who joined the 'executive-track' in a company in 2001, 65.1 per cent had already left the job (cf. 29.2 per cent among men). The proportion of women who are now Section Chief or Manager accounts for 12.4 per cent, whereas that to men is 39.2 per cent ([24]). Even in 2010, the dual-career track system was the predominant employment system in many firms and 49.2 per cent of large firms with more than 5,000 employees still use this.

In 2011, 13,373 women and 28,845 men applied for a job in the 'executive-track'. Women who were hired for this track accounted for 11.6 per cent (cf. men, 88.4 per cent). This means that only 1.6 per cent of women were successful in obtaining the post (cf. men, 5.8 per cent). When examining the 'general duties' position, 5,186 applicants were female, while only 369 applicants were male. Among those successful candidates, 86.0 per cent were women, and 14.0 per cent were men. The success rates for women and men are 5.8 and 13.3 per cent, respectively.

The natural consequence of this situation is that the ratio of women in managerial positions remains very low, as at the entrance point there are already fewer women on the executive track and many of them resign. All the data suggest that there are upper trends for all the scores, but the changes are rather slow and the sheer proportions are remarkably low.

By the end of twentieth century, women's place in Japanese society was barely maintaining a fine balance between home and workplace. Pressured by the international agenda of gender equality, domestic changes in politics and society created an environment where many family-friendly policies were introduced. However, these policies were primarily designed to solve the declining fertility rate in the country, in order to ameliorate the negative impact of more women working outside the home. Many stakeholders in this issue trod very cautiously,

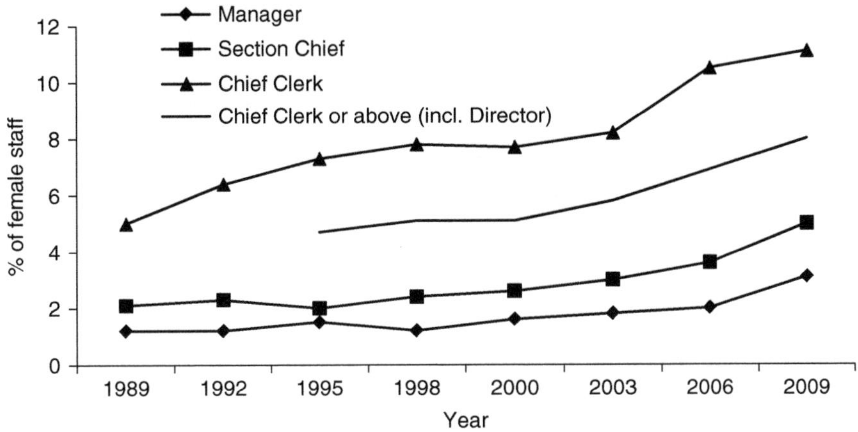

Figure 3.1 Ratio of women in firms by positions, 1989–2009.

Source: [32].

and did not readily embrace these changes. The 'equal participation of men and women', instead of 'gender equality', was the chosen phrase. This phrasing underscores the tenacity of traditional 'family' values in Japan, rather than the victory of mobilised social support for gender equality, especially outside the home. The principle of equal participation was therefore deemed to be the only realistic approach to maintain this balancing act.

The remainder of this chapter will examine what impacts these changes had on women's access to higher education and the labour market, and their choice of subjects during the 1990s onwards.

3.7 Changes in the meaning of education and profession for women in STEM

The year 2011 was the 25th anniversary of the EEOL. The ageing population had steadily increased, and the entire population began showing signs of decline. Amid these trends, the entire female labour market participation was at its peak in 1998, and also started to decrease. However, by comparison with 1986, the female labour force had increased by approximately four million. The rate of female participation in the labour market had also grown.

The education at attainment of women is particularly remarkable, and in 2008, female university graduates numbered 240,000 (cf. 90,000 in 1985). Furthermore, out of those 240,000, 180,000 successfully obtained jobs. The proportion of successful employment therefore accounted for 74.6 per cent. Concerning male graduates, 210,000 out of 320,000 obtained employment in 2008, which means that 66.4 per cent was successful. Comparing these figures only, female graduates are now in a better position than men at the entry point to the labour market. The successful rates of getting a job after graduation from different types of education institutions are shown in the following two figures.

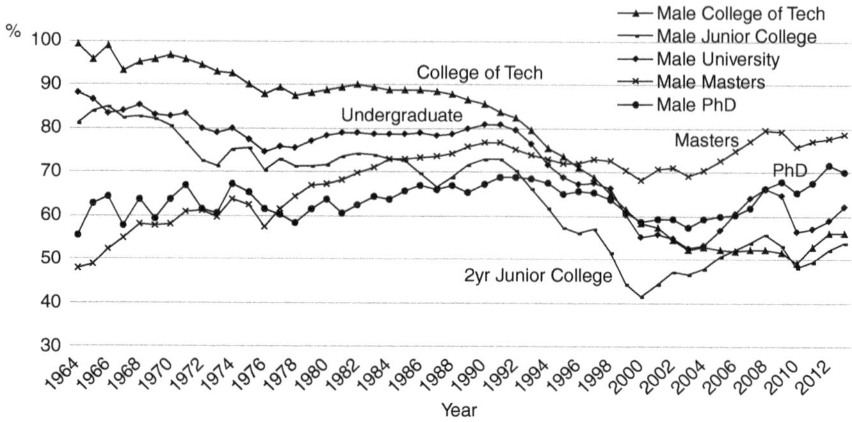

Figure 3.2 Success rate of male students securing a job after graduation, by type of educational institution, 1964–2013.

Source: [33].

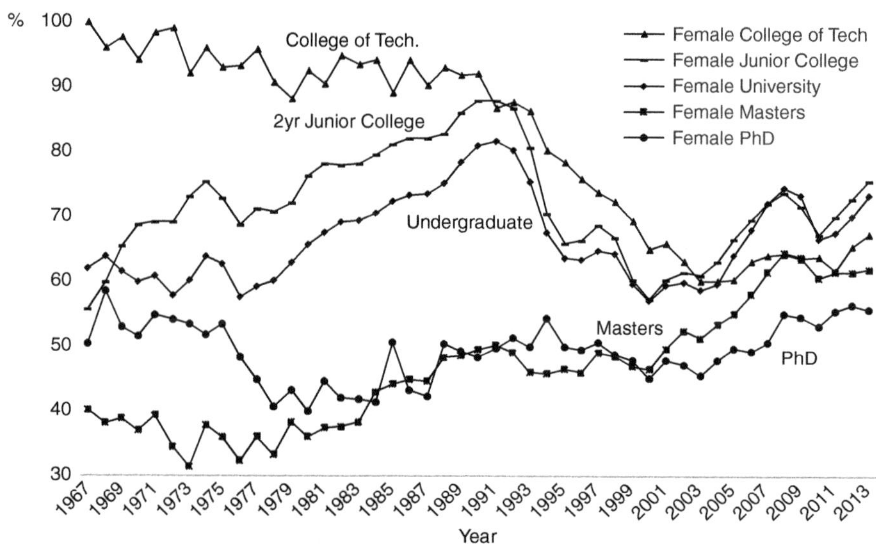

Figure 3.3 Success rate of female students securing a job after graduation, by type of educational institution, 1967–2013.

Source: [33].

When comparing the entry rates to the labour market for men after graduation from tertiary education, graduates from universities, junior colleges and colleges of technology all fail to maintain the same levels of success. This also applies to those with PhD degrees to a lesser degree. However, graduates with a Master's degree escaped the economic downturn rather unscathed, with the success rates of 70 per cent.

When the figures are examined for women during the 1990s, higher rates of success for graduates from colleges of technology are remarkable, despite the declining trends. What is interesting is that from 2003 onwards, graduates from junior colleges and universities recovered rather quickly, and postgraduates trailed behind them. However, since 2001, success rates for women with a Master's degree have accelerated.

In general, the recession from the 1990s affected men's success rates in obtaining employment more severely than women's. But the different trends between men and women suggest that employers expect male graduates to attain postgraduate degrees, while they regard junior college or undergraduate degrees as sufficient credentials for women employees.

The figure below illustrates the M-shaped curve, which indicates that women in Japan normally work from graduation until marriage or pregnancy (i.e. high labour participation rates among women in their 20s, and low in their 30s). This trend is particularly pronounced in Japan and South Korea ([25], [35]).

As of 2007, in total, 2.3 million women with university degrees or higher are engaged in full-time employment. There are 510,000 women (22.0 per cent) in the education and learning support sector, followed by 360,000 (15.5 per cent) in the healthcare and welfare sector, and 330,000 (14.5 per cent) in the wholesale and retail sector. A significant difference can be found across generations. Among women over 40 years of age, with a university degree and in regular employment, more than 40 per cent are in the education sector (280,000, 41.4 per cent).

On the other hand, in the younger generation, the proportion of women in education and learning support is much lower (9.7 per cent of 20–24 year-olds, and 11.3 per cent of 25–29 year-olds). This clearly shows that the choice of employment for women has diversified. In their 20s, women in the wholesale and retail sector account for 15-20 per cent (20–24 years: 20.2 per cent, 25–29 years: 15.7 per cent),

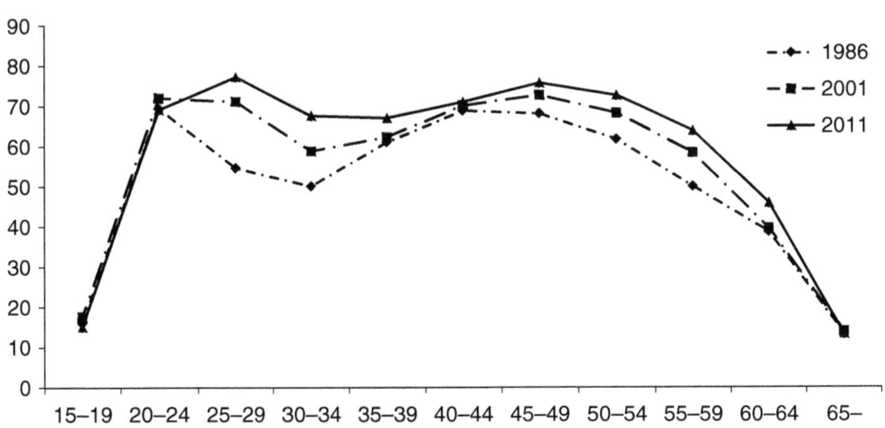

Figure 3.4 Female labour participation rates in 1986, 2001 and 2011.

Source: [32].

while women in the healthcare and welfare sector account for approximately 15 per cent (20–24 years: 15.5 per cent, 25–29 years: 16.6 per cent). Roughly 10 per cent of those in the finance and insurance sector are women (20–24 years old: 8.1 per cent, 25–29 years old: 14.0 per cent). Women in the education sector are mostly working as teachers, particularly among older age groups.

Out of 2.3 million women with a university degree or higher and in full-time employment, 960,000 are 'professional and technical workers' (41.8 per cent), and 'clerical workers' (39.1 per cent).

Clerical workers are the largest group among women in part-time employment (480,000, 41.4 per cent) and are followed by professional and technical workers (320,000, 27.4 per cent). There are 100,000 teachers, which accounts for 8.8 per cent.

In both full-time and part-time categories, the proportion of 'clerical workers' is higher than 'professional and technical workers' in younger age groups (20–30 years of age).

Comparing students' choice of majors at the university level between 1986 and 2008, social sciences and health have increased in popularity among women while humanities and education have decreased. For men, not much change is observable; the combined proportion of social science and engineering accounts for approximately 70 per cent.

As shown in Table 3.1, the ratio of women specialising in sciences decreased from 2.6 per cent in 1986 to 2.0 per cent in 2008, while that of women in engineering increased from 2.3 per cent to 4.2 per cent over that period. Yet in engineering, the gap between men and women specialising in engineering (4.2 and 24.8 per cent, respectively) is still great. In this sense, women enrolled in the College of Technology and in various universities' engineering faculties are still rare. The next section will look at the experiences of three generations of female graduates from the College of Technology, who have worked as professionals in the field.

Table 3.1 Choice of specialisation at university by gender (1986 and 2008 compared)

	Female %		*Male %*	
	1986	*2008*	*1986*	*2008*
Humanities	35.4	25.2	7.6	8.9
Social Sciences	15.1	27.6	46.1	41.5
Sciences	2.6	2	3.7	4.2
Engineering	2.3	4.2	25.3	24.8
Agriculture	2.1	2.9	3.9	3
Health	9.5	12.3	5.9	6.7
Home Econ.	7.7	5.7	0	0.4
Education	16.9	8.8	4.9	4.2
Art	6.9	5	1.2	1.5
Others	1.5	6.3	1.4	4.8
Total	**100**	**100**	**100**	**100**

(*Source*: [33].)

3.8 Survey results from Colleges of Technology

Colleges of Technology in Japan were established in 1962, with the aim of fostering talents and educating skills in science and engineering. After the enactment of EEOL in 1986, female applicants for these colleges increased, as mentioned in Chapter 2.

After the year 2000, Colleges of Technology across the country responded to a call for supporting gender equality, and distributed surveys targeting female engineers and graduates from the Colleges.

This section focuses on one of those surveys entitled 'Research report on the career formation process of female engineers' ([26]). The survey was conducted between 2007 and 2009, by college lecturers, with a focus on differences in the way female engineers work depending on generation.

Survey participants were female graduates from Colleges of Technology, the majority of whom are currently engaged in employment. The age range is 30–64. Out of the 64 respondents, 57 women work in technical positions today, five have changed their type of job (no longer technical workers), and two are no longer employed (currently unemployed).

As a result, the participants were divided into two groups: 51 graduates from Colleges of Technology, and six graduates who pursued further degrees.

In terms of types of affiliations, the majority of graduates work in firms (40 persons). Eleven women are working either as self-employed or in the design office. There are also four public officials, and two working in the education sector.

The survey data were analysed by splitting the participants into three 'generations'. The first generation (45 years of age or over) started to work before the EEOL was enacted in 1986. The second generation (between 40 and 44 years of age)

Table 3.2 Number of survey participants by generation

		30s	*40s*	*50s*	*60s*	*Total*
Mechanical Eng.	College of Technology			1		1
	Further degree					0
Electrical Eng.	College of Technology	3	1	1		5
	Further degree					0
Electronics	College of Technology	1				1
	Further degree	2				2
Civil Eng.	College of Technology	2	1		1	4
	Further degree					0
Architecture	College of Technology	4	5	2		11
	Further degree					0
Chemistry & Biochemistry	College of Technology	7	5			12
	Further degree	1				1
Information & Telecommunications	College of Technology	14	2			16
	Further degree	2				2
Shipping Tech.	College of Technology	1				1
	Further degree	1				1
Total		**38**	**14**	**4**	**1**	**57**

(*Source*: [26]: 6. Adapted by the authors.)

Table 3.3 Number of survey participants by types of job

	College of Technology	*Further degree*	*Total*
Research & Development	1	1	2
Development & Design	21	4	25
Systems Engineer	4	0	4
Production & Quality Control	1	0	1
Operation Support & Management	1	0	1
Service Engineer	2	0	2
Sales	2	0	2
Captain	1	0	1
Supervisor & Director	0	1	1
Environmental management	3	0	3
Executive-career track	4	0	4
Professional & technical workers	4	0	4
Clerical workers	3	0	3
Others	4	0	4
Total	**51**	**6**	**57**

(*Source*: [26]: 7. Adapted by the authors.)

graduated between 1986 and 1991 when economic growth was at its peak. The third generation belongs to the post-bubble economy era.

> *First generation (45 years old or older): 8 persons*
> Frontier generation. They have experienced many things including the guarantee of parental leave for the first time as women. The ratio of female students at Colleges of Technology was as low as the 1–3 per cent range, and consequently there were no facilities such as women's restrooms.
>
> *Second generation (40–44 years old): 11 persons*
> The generation that entered the workforce after the EEOL's passage and during Japan's bubble economy. The number of job offers increased, and women did not have much difficulty finding a job. This generation weathered a 'restructuring' period (with many layoffs). They sought to manage a better balance between work and family as society began to improve childcare facilities and companies implemented family-friendly policies.
>
> 'When I graduated, it was in the heyday of the bubble. A College of Technology student specialising in IT was offered a job by 100 to 200 companies'. ([26]: 21)
>
> The female students' enrolment ratio soared from 3.7 per cent in 1985 to 7.7 per cent in 1990, when Colleges of Technology founded the Department of Information and Telecommunications.
>
> *Third generation (30–39 years old): 38 persons*
> The generation that entered the workforce in the post-bubble economy. They managed to obtain jobs at a time of economic downturn, the so-called

'employment ice age'. Since the BAGES and the revised EEOL were enacted in 1999, women's social advancement was promoted through legislation.

> 'One year after we graduated, the bubble burst. If we had gone to a university (instead of the College of Technology), we would not have managed to get a job' (chemical companies, career-track, 30s). ([26]: 22)
>
> 'It was a really tough period for women to get a job … some firms rejected my application because I am a woman …' (IT, development and design, 30s). ([26]: 22)

The rates of successful entry into the labour market confirm these comments. The highest points for graduates from both Colleges of Technology and universities was in 1992 (87.6 per cent for Colleges of Technology, and 80.4 per cent for university graduates). In ten years, the rates have dropped to 60.1 per cent for Colleges of Technology graduates, and 58.5 per cent for university graduates ([33]).

Some caution is required when interpreting these findings because the figure representing successful jobseekers as a percentage of the total number of graduates also includes those who did not seek jobs. These numbers therefore can be misleading. In addition, this overlaps with the time when more students began pursuing further degrees.

Prior to this period, the major type of employment for female graduates was a mineral-based chemical company. However, the majority of women entered IT-related (including electronics and communication) companies.

This study also analyses differences in attitudes towards work and family life among the three generations. There is a great difference in the retention rate for women in their first job: between 28.6 per cent in the 40s and 76.3 per cent (78.1 per cent, if those who went on to study further are included) in the 30s. All of the first generation experienced a career change. The reasons were all related to 'family' (e.g. marriage, baby birth, relocation of spouse, and dependents and care of the family).

> 'Someone I know, a woman who graduated with a degree in Mathematics from a national university, had to give up her job because of her husband's work transfer … many had no choice but to quit. Among talented people, there are not many who continued to work. It was just the normal thing to do …' (Educational institutions, 60s). ([26]: 23)

The woman referred to here changed her job four times in the course of her career, and three out of those four times the reason was her husband's work transfer. For this generation, five out of eight people ended up becoming 'self-employed' after several changes.

Regarding the second generation, the picture is quite different. Seven out of 11 people experienced a job change, while four continued in their original job. The reasons for job change vary as well. In addition to family-related issues, they included personal health issues, busyness, and layoffs due to the company's decline in business.

This generation considers continuing work as nothing unusual. There were some examples of headhunting through networking and promotions based on the recognition of skills. The advantage of having strong expertise and professional skills began to pay off in some cases.

However, some conventions within firms, such as resignation after marriage and disparities between men and women in wages and working conditions, were still present.

> 'Women were not allowed to get onto the career track, but my manager at that time negotiated extremely hard with Human Resources, making the case that I should be allowed to get onto the track. Thanks to him, I was able to become a "professional" worker' (Electric companies, Design and Development, 40s). ([26]: 24)

In addition, the general perception of society back then was that 'women should stay at home to care for family'. Criticism aimed at this generation of working, professional women was very severe.

> 'When I told people that I have children but am also working, everyone said "your husband must be very understanding", and "your company must be very lenient". But I always thought "what about me? I am the one who is working hard here, am I not? Who can actually recognise all my effort?" Society thinks totally differently about the whole thing. I thought because I want to work, it does not matter whether my husband or company are unsupportive … (women's right to work is legally protected)' (chemical enterprise, management and audit, 40s). ([26]: 25)

Among the third generation, only eight out of 38 have changed jobs. The rest (30 women) kept their original job. Concerning a possible future job change, participants mentioned family-related issues (baby birth and long-term care for their elderly). Some mentioned the 'exit after marriage' route as a possibility, while others talked about the career change as a potential advancement in their professional career, which was different from the first and second generation. Many of them received support from their own or their partners' parents in rearing children.

As the 2009 White Paper on Gender Equality indicates, by 1992, the number of double-income households surpassed that of single-income households. Twenty years have passed, and the idea of working women is no longer considered 'abnormal'. Yet, the form and style of working varies greatly between generations.

Looking at the current position and its relationship with years of employment, it becomes clear that the longer an employee stays in the same firm, the likelier is the woman to be promoted to a senior position. According to the Labour

Survey ([27]), the major reasons for the small number of women in senior positions were as follows.

- 'No women in my firm who have the knowledge, experience, and judgment necessary for the post' (46.9 per cent).
- 'As length of their service is too short, they retire before they become managers' (30.9 per cent).
- 'There are women who may take up the management post in the future, but at the moment, no one meets the criteria in terms of length of working years in order to hold a senior position' (27.9 per cent).

The second generation, in particular, may have entered the labour market, when the career-track system was not applied at all to women (i.e. prior to stricter enforcement of the revised EEOL).

> 'When I joined the company, the company was still allowed to have different training programmes for male and female employees' (chemical companies, management and audit, 40s). ([26]: 29)
>
> 'Because there was no precedent, women were not taken into account, not considered for the career track' (electricity companies, development and design, 40s). ([26]: 29)

However, for the third generation, the career-track path became open to both men and women working for leading manufacturers. One person who is not so career-driven was promoted to a management position in her early 30s (IT systems, research and development, Master's degree), while in another case, one was unable to further her career, because her final degree (from a College of Technology) became a 'reason' that went against her promotion (chemical enterprise, planning and research, also in her 30s) ([26]: 29). Among the third generation, some are satisfied with the status quo and thus exhibit little career ambition.

The most common job type is 'design and development'. Most of the first and second generation, are self-employed, in which they take on the role of 'design and development'. The third generation is assigned to 'design and development' in their first job. In addition, in the third generation, the effect of social change after the EEOL can be observed. A higher-level degree seems to be required for the third generation in 'design and development'. Some firms stopped recruiting and hiring graduates from Colleges of Technology ([26]: 31).

> 'Because companies only employ university graduates, I decided to pursue further studies' (chemical companies, design and development, 30s) ([26]: 33).
>
> 'I heard that in some companies, the type of job you are asked to do depends on your final degree. There is an invisible line that separates technical college

graduates and university graduates. So no matter how hard you work as a technical college graduate, the line clearly determines your type of work and you cannot cross that line … I've heard that a project-team leader has to be a university graduate, while a technical college graduate will always serve as an assistant' (chemical companies, design and development, 30s) ([26]: 33).

Aspirations and overall demand for higher degrees can be observed among technical college students, particularly those who would like to pursue a career in 'design and development' and 'research and development'.

3.9 Female researchers in STEM: where are they?

Compared with other fields, the ratio of women in research is still relatively low. In the legal professions, 17.7 per cent of judges and 17.5 per cent of lawyers are female. Of those who passed the bar exam in 2012, the proportion of women was 25.9 per cent ([28]). In the civil service, out of those who are hired by ministries through a highly competitive exam, 28.6 per cent are women. In medicine, 18.9 per cent of medical doctors and 20.8 per cent of dentists are women ([29]).

Researchers are in the minority even within the entire female population, although the ratio has seen a steady increase. In 2006, it was reported that the number of female researchers across all disciplines has exceeded 100,000 for the first time (102,900 out of 830,000 in total). By affiliation, 21.5 per cent of all researchers (63,400 people) are affiliated with universities, while 6.5 per cent (33,800 people) are working in companies (Figure 3.5).

The total number of researchers in Japan is 893,000, of whom 14.0 per cent are female (as of 2012). Since universities have the highest proportion of female researchers, this reflects the striking gap between men and women in university

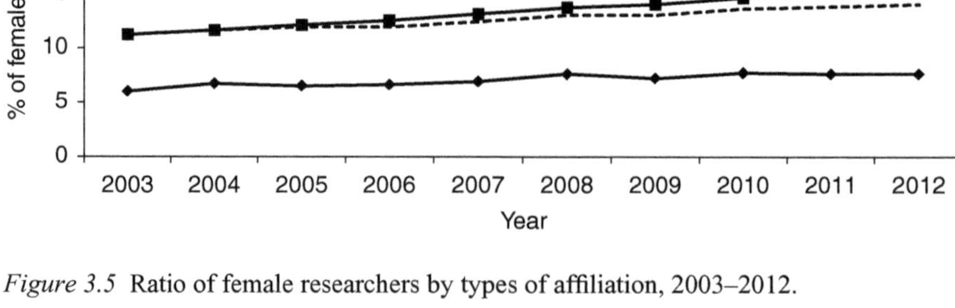

Figure 3.5 Ratio of female researchers by types of affiliation, 2003–2012.

Source: [32].

posts. Only four per cent of all presidents are women, followed by seven per cent (Vice President), 14 per cent (Professor), and 22 per cent (Associate Professor). Female lecturers account for almost 30 per cent, as of 2013.

This trend of fewer women in higher positions is applicable to all disciplines (see Figure 3.6). Given the low proportion of female students specialising in sciences and engineering in the first place (see Figure I.1 in Introduction), it is only natural that among all the female researchers, less than 10 per cent of them are professors and associate professors in sciences, engineering and agriculture. In sciences and agriculture, more than half of assistant researchers are female,

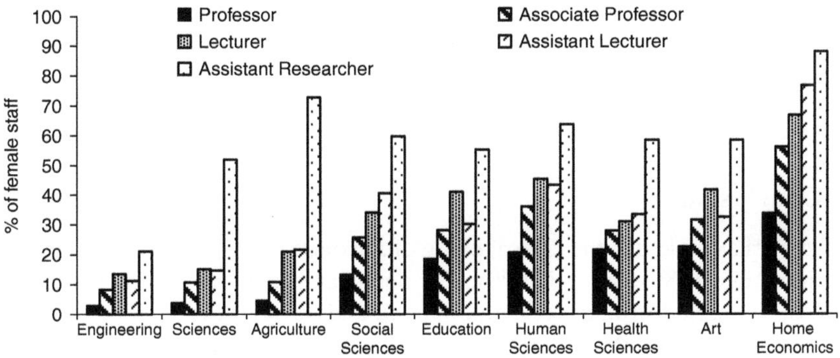

Figure 3.6 Ratio of female academics in STEM fields.

Source: [32].

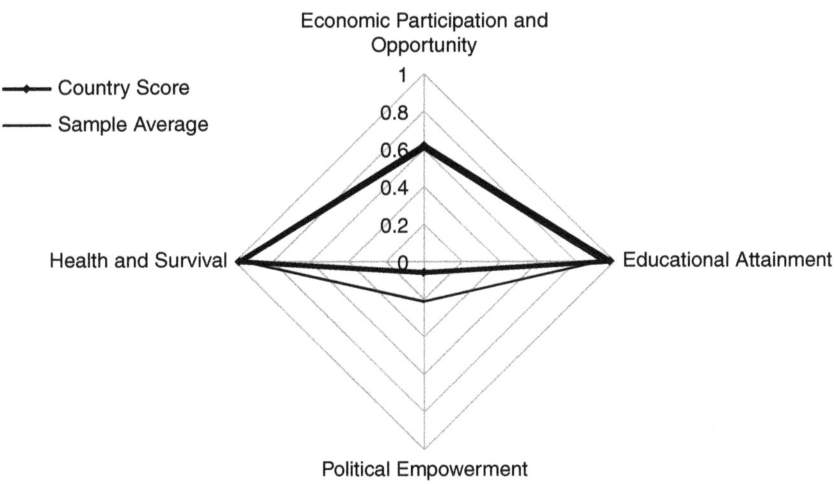

Figure 3.7 Gender Gap Index for Japan, 2014 (0.00=inequality, 1.00=equality).

Source: [31]: 220. Adapted by the authors.

just as in other disciplines. However, in engineering, the ratio of female assistant researchers reaches only 20 per cent ([30]).

On the other hand, education for both men and women has always been taken seriously. When the Human Development Index in (HDI) is used, a high score is given to the Capacity Development Indicators for women in Japan. After adjustment is made for gender inequality (GDI), Japan ranks 13 out of 144 countries. This indicates a relatively high level of education for Japanese women in general. However, an examination the World Economic Forum's Global Gender Gap Index (GGI) reveals that the proportion of women in leadership positions is extremely small; Japan ranked 104 out of 142 countries in 2014. Since Japan's GGI score (0.658) has lagged behind other countries for so many years, the situation surrounding women in STEM represented this trend (see Figure 3.7).

The 2014 Report explicitly referred to this fact by stating 'Japan is among the countries with the biggest differences between the percentage of female and male students enrolled and graduating in STEM studies' ([31]: 27). This is reflected in Figure 3.8.

3.10 Conclusion

This chapter has looked at the critical phase of Japanese political and social history between the late 1980s and the 1990s. The post-war system began unravelling at the end of the period of high economic growth and the start of the period of low population growth. While concepts such as 'gender equality' and 'gender free society' were still resisted and eschewed, 'equal participation of men and women' in society gradually made its way into the main political discourse. With additional international pressure exerted at around the 4th World Conference on Women in 1995, policy changes ensued, culminating in the enactment of the BLGES in 1999.

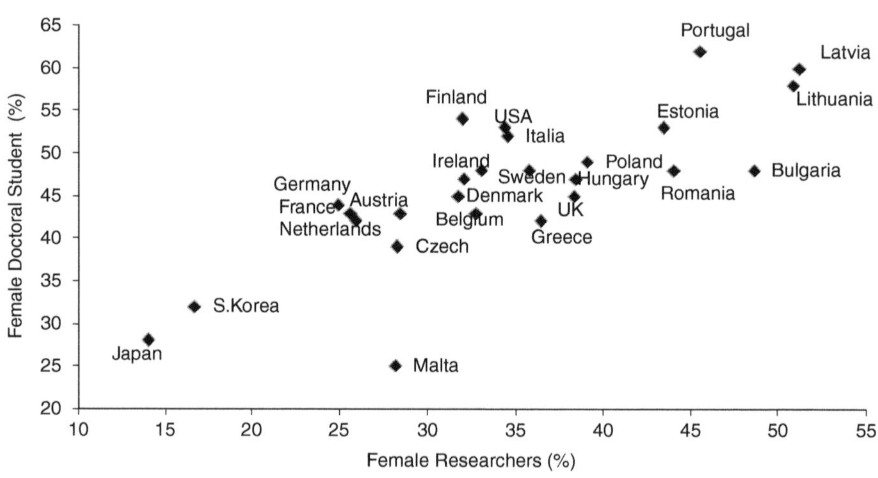

Figure 3.8 Proportion of female researchers and doctoral students, 2013.

Sources: European Commission 2012; UNESCO 2013.

This chapter has also highlighted a slow but steady increase in the number of women entering higher education (i.e. a transition from two-year colleges to a four-year universities) and the labour market. Despite the slow progress in gender equality in STEM domains, the Colleges of Technology founded in 1962 welcomed more women in science and technology after the EEOL. Their work experiences after graduation were compared and contrasted across three different generations.

By the mid-2000s, it became clear that Japan had to do something in response to a rapidly ageing population, combined with the relative decline in the international competitiveness of its economy. A systemic overhaul is considered necessary, and however recalcitrant, a recalibration of Japanese society and the policy-making structure will occur. These changes finally paved the way to drastic policy formulations towards positive action in science and engineering. The next chapter will outline how policy changes took place and were implemented.

Notes

1 The Japan Socialist Party (Nihon Shakai-tō) became the Social Democratic Party of Japan (Shakai Minshu-tō) in January 1996.
2 Akiko Dōmoto is a former journalist, alumna of TWCU, and subsequently became the first female governor of Chiba Prefecture, 2001–2009, the third female Prefectural Governor in Japanese history.
3 Professor Sumiko Iwao is Emeritus Professor of Social Psychology at Keiō University and an alumna of Keiō and Yale Universities.

References

[1] B. Bishop. *Globalisation and Women in the Japanese Workforce*. Abingdon, Oxon: Routledge, 2004.
[2] M. Nakano. *The Policy-Making Process in Contemporary Japan*. London: Macmillan and St. Martin's Press, 1997.
[3] L.J. Schoppa. Zoku power and LDP power: A case study of the zoku role in education policy. *Journal of Japanese Studies*, 17: 79–106, 1991.
[4] S. Takahashi. The political dynamics of government change. *The University of Tokyo Journal of Law and Politics*, 7: 1–28, Spring 2010.
[5] J.A.A. Stockwin. *Governing Japan: Divided Politics in A Resurgent Economy*, chapter on Who runs Japan?, pages 135–155. Malden, MA: Blackwell Publishing, fourth edition, 2008.
[6] N. Hiwatari. The politics of reform and transformation of the postwar party systems in Japan and Italy. *The Journal of Social Science*, 50(2): 99–126, 1999.
[7] T. Kashima. *Danjo kyōdō sankaku no jidai* (Era of equal participation of men and women). [In Japanese] Tokyo: Iwanami Shinsho, 2003.
[8] M. Osawa and C. Ueno (ed.) *Radikaru ni Katareba (Radically Speaking)*, chapter 'Danjo kyōdō sankaku kihonhō no mezasu mono: sakutei made no omote to ura' (What the Basic Law of Gender Equality aims to achieve: behind the scenes of making the law) [In Japanese]. Tokyo: Heibonsha Publishing, 2001.
[9] M. Daly. *The Oxford Handbook of the Welfare State*, chapter 9: 'Families versus state and market', pages 139–151. Oxford: Oxford University Press, 2010, eds. F. Castles, S. Leibfried, J. Lewis, H. Obinger, and C. Pierson.

[10] J. Gottfried and J. O'Reilly. Re-regulating breadwinner models in socially conservative welfare systems: Comparing Germany and Japan. *Social Politics: International Studies in Gender, State and Society*, 9(1): 29–59, 2002.

[11] I. Holliday. Productivist welfare capitalism: social policy in East Asia. *Political Studies*, (48): 706–723, 2000.

[12] C. Aspalter. The East Asian welfare model. *International Journal of Social Welfare*, 15(4): 290–301, 2006.

[13] Y. Lee and Y. Ku. East Asian welfare regimes: testing the hypothesis of the developmental welfare state. *Social Policy & Administration*, 41(2): 197–212, 2007.

[14] F.G. Castle. The world turned upside down: Below replacement fertility, changing preferences, and family-friendly public policy in 21 OECD countries. *Journal of European Social Policy*, 13(3): 209–228, 2003.

[15] T. Fahey. Familism and fertility: Ireland from a Japanese point of view. *The University of Tokyo Law and Politics Journal*, 11: 89–107, 2014.

[16] P. Boling and F.M. Rosenbluth (eds). *The Political Economy of Japan's Low Fertility*, chapter on 'Policies to support working mothers and children in Japan', pages 131–154. Stanford University Press, 2007.

[17] A.K. Abe. Effects of child-relating benefits in Japan. *Journal of Population and Social Security: Social Security Study* (August), 2002.

[18] C. Fujiwara. *Danjyo kyōdō sankaku shakai to shimin (Equal Participation of Men and Women in Society and Citizens).* [In Japanese] Tokyo: Musashino University Press, 2012.

[19] E. Shinotsuka. *Women of Japan and Korea*, chapter on Women Workers in Japan: Past, Present, Future. Philadelphia, Temple University Press, 1994, eds. Joyce Gelb and Marian Lief Palley.

[20] M. L. Starich. The 2006 revisions to Japan's equal opportunity employment law: a narrow approach to a pervasive problem. *Pacific Rim Law & Policy Journal*, 16(2): 551–578, 2007.

[21] A. Shibata. *Women's Work in the World Economy*, chapter on The effects of Japanese income tax provisions on women's labour force participation. New York: NYU Press, 1992. Nancy Folbre, Barbara Bergmann, Bina Agarwal, and Maria Floro (eds).

[22] S. Shiota. *Nihon no shakai seisaku to jendā: danjo byōdō no keizai kiban (Japan's Social Policy and Gender: Economic Basis of Gender Equality).* [In Japanese] Tokyo: Nihon Hyōronsha, 1997.

[23] H. Takeda. *The Political Economy of Reproduction in Japan: Between Nation-State and Everyday Life*. Abingdon, Oxon: Routledge, 2005.

[24] Ministry of Health, Labour and Welfare. Kōsu betsu koyō kanri seido no jisshi/shidō jyōkyō (Investigation into the dual-track career system) [In Japanese]. [online] Available from: http://www.mhlw.go.jp/general/seido/koyou/danjokintou/dl/cource_joukyou.pdf, 2013. Accessed 15 November 2014.

[25] OECD. *Babies and Bosses – Reconciling Work and Family Life: A Synthesis of Findings for OECD Countries*. Paris: OECD Publishing, 2007.

[26] Y. Uchida *et al.* Josei gijutsusha no career keisei katei ni kansuru jisshō kenkyū: kōsen no kyōku model kōchiku wo mezashite (Research on career development process of female engineers: building a model for technical college education) [In Japanese]. Technical report, Grant-in-Aid for Scientific Research Basic Research (C-1) Final Report for JSPS, 2010.

[27] Ministry of Health, Labour and Welfare. Daisotsu jyosei no hataraki kata (The Way Female University Graduates Work). [In Japanese]. [online]. Available from: http://www.mhlw.go.jp/houdou/2009/03/dl/h0326-1d.pdf, 2009. Accessed 12 February 2014.

[28] Gender Equality Bureau, Cabinet Office. Danjyo kyōdō sankaku shakai no keisei no jyōkyō (Status Quo and Developments of Equal Participation of Men and Women) [In Japanese]. [online]. Available from: http://www.gender.go.jp/about_danjo/whitepaper/h25/zentai/html/zuhyo/zuhyo01-01-14.html, 2013. Accessed 25 November 2014.

[29] Gender Equality Bureau, Cabinet Office. Kaku bunya ni okeru shidōteki chii nado ni shimeru jyosei no wariai (Proportion of Women in Leading Positions in Various Fields) [In Japanese]. [online]. Available from: http://www.gender.go.jp/policy/positive_act/pdf/positive_action_002.pdf (2012). Accessed: 25 November 2014.

[30] M. Kato, H. Chayama, and A. Hoshikoshi. Nihon no daigaku kyōin no jyosei hiritsu ni kansuru bunseki (Analysis of Female Staff Ratio in Japanese Universities). [In Japanese]. Technical report, National Institute of Science and Technology Policy, 2012. [online]. Available from: http://www.nistep.go.jp/achiev/ftx/jpn/mat209j/pdf/mat209j.pdf Accessed 25 November 2014.

[31] World Economic Forum. Technical report, Geneva: World Economic Forum. [online]. Available from: http://www3.weforum.org/docs/GGGR14/GGGR_CompleteReport_2014.pdf, 2014. Accessed 9 November 2014.

[32] Ministry of Health, Labour and Welfare. *Heisei 22 nendo koyō kintō kihon chōsa* (Basic research into Equal Employment Opportunity 2010). [In Japanese] [online]. Available from: http://www.mhlw.go.jp/toukei/list/71-22.html, 2011. Accessed 15 November 2014.

[33] Ministry of Education, Sports, Science and Technology. *Gakkō kihon chōsa* (School basic survey) [In Japanese] [online]. Available from: http://www.mext.go.jp/b_menu/toukei/chousa01/kihon/1267995.htm, various years. Accessed 15 November 2014.

[34] European Commission. She Figures 2012: Gender in Research and Innovation. Technical report, Brussels: DirectorGeneral for Research, 2012. [online]. Available from: http://ec.europa.eu/research/science-society/document_library/pdf_06/shefigures-2012_en.pdf, 2013. Accessed 9 November 2014.

[35] K. Sechiyama. *Higashi ajia no kafuchōsei – jendá no hikaku shakaigaku* (Patriarchy in East Asia: A comparative sociology of gender [In Japanese]. Tokyo: Keisō Shobō, 1996.

4 Policy advances in the promotion of equal participation of men and women, late 1990s and 2000s

4.1 Introduction – Reforming central government

As noted in the previous chapter, one of the major obstacles that prevented policy change in the Japanese political system was a compartmentalised public administration system, geared towards serving and defending vested interests in the iron triangle.

When the coalition government of three parties (LDP, Japan Socialist Party and Harbinger Party) collapsed in the autumn of 1996, Ryūtarō Hashimoto called for a general election to restore the LDP-led majority in the Diet. The Hashimoto government (1st: January–November 1996; 2nd November 1996–July 1998) tried to seize the opportunity, embarking on a series of reforms to break the iron triangle, later continued by the Koizumi government in the early 2000s (see [1] for the Hashimoto reform and [2] [3] for the Koizumi reform).

Hashimoto returned as Prime Minister of a single minority government with an informal collaboration with the two parties. To win back public support, Hashimoto developed an ambitious reform package, including VAT tax reform (increasing the tax rate from 3 per cent to 5 per cent), and six major reforms (fiscal reconstruction, financial deregulation, social security, education, decentralisation and public administration). Among these reform pledges, public administration reform was one of the key agendas to increase public popularity, as corruption within the civil service had been causing public distrust since the early 1990s. Hashimoto aimed to transform the government, weaken bureaucratic control and strengthen the power of the political parties and the prime minister ([4]).

As soon as he became prime minister, he set up the Administrative Reform Council (ARC) and took the chairmanship, inviting big business leaders, journalists and academics onto the Council. Chaired by the prime minister himself, the ARC after November 1996 looked to British practices of agencification, and decided to emulate them and rationalise national universities and hospitals as part of public administrative reform ([5]). The concept was to divide every central ministry into two functions (policy-making and execution) in order to slim down the whole organisation while enhancing performance and productivity. The ARC proposed that these mechanisms should be applied not only to hospitals and schools, but also to museums, national parks, and the meteorological and aviation agencies, covering 56 agencies in total. This effort represented a comprehensive restructuring of central administration, covering the whole range of policy domains and therefore involving whole sections of the ministries. The Basic Act for Reforming Central Administration was tabled in the Diet on 17 February 1998. The central government ministries were radically reorganised in 2001. Many public bodies were also converted into Independent Administrative Institutions (IAIs, *Dokuritsu Gyōsei Hōjin*). Management methods of private sector corporations were emulated, and IAIs were given considerable autonomy in deciding the budgets and operations. All the national universities became IAIs, and the MEXT-affiliated body, National Women's Education Center (NWEC), was re-established as an IAI in 2001.

This reorganisation effort had a major impact on women in science and technology. Three things are worth noting.

First, the Council for Science and Technology Policy (CSTP, *Sōgō Kagaku Gijyutsu Kaigi*) was newly created under the control of the Cabinet Office as the nation's highest science advisory body. This reform meant that the roles of and selection processes for the Science Council of Japan (SCJ) had to be reviewed. It was later reorganised in 2005. As noted in Chapter 2, the number of SCJ's female members remained stable, between one and three, from 1981 to 2000. In 2000, six were selected across all disciplines, of whom two were in the natural sciences[1]. In 2003, the number doubled from six to 13, and four scientists came on board. When the reformed SCJ was established in 2005, the total female members increased to 42 (20 per cent), of whom 17 were in STEM.

Second, the major reorganisation of national universities prompted a debate concerning the future of Ochanomizu University as a women-only institution. A merger of Ochanomizu University with other co-educational institutions was considered on the basis that a publicly funded university should not discriminate against men now that the principle of equal participation was the law. However, this suggestion met with fierce opposition from various stakeholders. Their argument was that Ochanomizu University had produced so many prominent female leaders in the past, that this could not be matched by any other institution in Japan. Slow progress in achieving the goal of fostering more female researchers also helped the argument ([6]). In a very crowded and competitive education sector in Japan with so many universities, the governing authority of the university judged in the end that it could not afford to lose its unique feature as a women's university with such a good reputation and rich history. In turn, the government also began to recognise the need to support women's universities, rather than transform them into co-educational institutions ([7]).

Third, at the same time as the central government reorganisation occurred, the Office of Gender Equality was upgraded to the Gender Equality Bureau (*Danjo Kyōdō Sankaku Kyoku*) within the Cabinet Office, with responsibility for planning and coordinating the gender equality policies of the government. The Council for Gender Equality, chaired by the Chief Cabinet Secretary, was given the responsibility of monitoring the implementation of all the gender equality policies of the government. In May 2000, the Gender Equality Bureau published its first White Paper on Gender Equality (*Danjo Kyōdō Sankaku Hakusho*).

The ministerial post for Gender Equality (In Japanese, *Danjo Kyōdō Sankaku Daijin*) was also created, and Yasuo Fukuda became the first Minister (January 2003) under the Mori and Koizumi administrations.

4.2 Uniting forces of scientific societies: the establishment of the EPMEWSE

The Basic Act for Gender Equal Society (BAGES) enacted in June 1999 clearly set out the equal participation of men and women in society as the major goal. The new EEOL and the Labour Standards Act were both designed to help achieve that goal. Soon after the enactment of the BAGES, academic societies began playing an active role. In January 2001, the Japan Society of Applied Physics (JSAP) agreed that the lack of women in the field was an acute problem, and it needed to tackle it head-on. The Society has 25,000 members, 60 per cent of whom are researchers and engineers in private enterprises, 30 per cent from universities, and the rest from research institutions. At that time, only four per cent were women. The then Vice-President (Toshio Gotō (male), Nagoya University) requested that more female members be appointed as committee members. In order to identify potential candidates, a network of women physicists and engineers was formed in February 2001. The Society's branch, the Japan Optical Society, already had a forum called Contemporary Optics, consisting of women physicists and male advisors specialising in optics. Since the core members of this group were women, it formed the basis for the women's network in the Society.

Initially, there was resistance, scepticism and opposition to this initiative, particularly around positive action towards women. However after much deliberation, the network came to the conclusion that candidates should include both men and women, given the increasingly serious situation regarding the shortage of talent in science and engineering, high-tech industries, and wider society in Japan. The situation was exacerbated by a combination of the low fertility rate and the dwindling popularity of science and technology among young students. With the goal of changing the overwhelmingly negative perception about women physicists among the members, and promoting equal participation of men and women, the Society's Board of Directors decided to establish the Committee for Promoting Equal Participation of Men and Women (Chair: Kashiko Kodate (female, JWU), Vice-Chair: Yoshikazu Tōyama (male, Fujitsū)) in July 2001. The composition of the membership was 13 women and six men. The JSAP became the very first academic society to found such a committee.

At around the same time, the International Union of Pure and Applied Physics was organising its very first International Conference on Women in Physics in Paris, between 7 and 9 March 2002. The Physical Society of Japan (PSJ) and the JSAP decided to send a delegation. Initially, the two societies discussed and attempted to conduct a survey jointly, but the differences in opinion between scientists and engineers were so great that they carried out a survey separately.

Prompted by this failed attempt, yet successful joint attendance at the international conference, the JPS and the JSAP approached another large academic society, the Chemical Society of Japan (CSJ), to form a cross-society network association. As a result, the Japan Inter-Society Liaison Association Committee for Promoting Equal Participation of Men and Women in Science and Engineering (EPMEWSE) was founded in July 2002.

Various academic societies began highlighting the underrepresentation of women in STEM. In October 2002, the first EPMEWSE meeting was held. Full membership was given to 15 societies and observer status to 13 societies ([8]).

The establishment of a cross-society network was unprecedented, and provided the momentum to carry out a large-scale survey across the fields of science and engineering. As a founding member of EPMEWSE, the JSAP took the lead in conducting the survey report commissioned by the MEXT in 2003 as a research project under the auspices of the Lifelong Learning Policy Office. The survey received almost 19,291 responses (16.1 per cent of which were from women). The Chairperson of the EPMEWSE, Masako Bandō[2], wrote that 'the survey results elucidate the day-to-day activities of people working in the science and engineering fields, such as their working environments and family lives' ([9]: 3).

The Planning and Execution Representative of the EPMEWSE, Kashiko Kodate, noted that:

> 'most of the previous surveys focused on female professionals, with the goal of raising their status to the same level as males. In contrast, this survey was conducted with the cooperation of 39 academic organisations that belong to EPMEWSE, and a uniform set of questions was asked to every member in a science or engineering profession. The purpose of the survey was to accurately understand the current situation, and to make recommendations regarding practical solutions and effective enlightening activities'. ([9]: 4)

The survey shed light on differences in the working environments of academic institutions and private companies. Interestingly, this collaboration between different academic associations was also promoted under the banner of the 'equal participation of men and women' rather than 'equality':

> 'an openly feminist approach is an obstacle rather than a facilitator for change in Japan and it is very important to have men on board to improve the working environment for women, as (the male-dominated) society's mentality needs to be altered' ([6]: 322).

The next section will look at the survey results, and recommendations, which prompted government policy change.

4.3 Survey findings, the 2005 White Paper and policy change

The member of respondents by gender and association is shown in Table 4.1.

Table 4.1 EPMEWSE 2004 survey respondents by gender and society

Association	*Number of respondents*	*Respondents by gender*			*% of female respondents*	*% of female members (if known)*
		Male	*Female*	*N/A*		
Fire Sci. Eng.	49	48	1	0	2.0%	
Automo. Eng.	159	153	6	0	3.8%	0.6
Chem. Eng.	842	804	35	3	4.2%	3.8
Precision Eng.	130	123	6	1	4.7%	
Mechanical Eng.	1,172	1,099	69	4	5.9%	1
Geotech.	455	427	28	0	6.2%	1.5
Magnetics	282	264	18	0	6.4%	3
Atomic Energy	402	372	30	0	7.5%	2.3
Image Info. TV	291	267	23	1	7.9%	1.7
Appl. Phys.	3,068	2,817	244	7	8.0%	
Metal	271	245	25	1	9.3%	
Elec. Info. Comm.	2,274	2,056	212	6	9.3%	2.3
Math. Bio.	20	18	2	0	10.0%	
Physics	2,492	2,234	252	6	10.1%	4.5
Liq. Crystal	69	61	8	0	11.6%	
Iron Steel	161	142	19	0	11.8%	1.1
Mathematics	502	440	60	2	12.0%	5
Polymer Sci.	447	382	63	2	14.2%	7.1
Architecture	2,204	1,872	327	5	14.9%	9.2
Electrochem.	205	174	31	0	15.1%	
Chemistry	1,932	1,638	293	1	15.2%	8.4
Total	**1,9291**	**1,6140**	**3104**	**47**	**16.1%**	
Info. Process	659	547	110	2	16.7%	4
Protein Science	322	267	54	1	16.8%	
Biophysics	822	676	145	1	17.7%	14
Astronomy	259	211	46	2	17.9%	7.3
Others	5,072	4,124	938	10	18.5%	
Bio. Space	48	39	9	0	18.8%	12.8
SGEPSS	107	86	21	0	19.6%	
Comp. Endocr.	175	140	35	0	20.0%	14.9
Analytical Chemistry	178	142	36	0	20.2%	10.5
Physiology	370	288	82	0	22.2%	14.2
Biochemistry	1,329	998	329	2	24.8%	19.6
Zoology	584	427	155	2	26.6%	20.2
Illum. Eng.	67	49	18	0	26.9%	14.3
Mol. Biology	2,878	2,079	795	4	27.7%	20.1
Plant Physiology	499	354	145	0	29.1%	18.7
Developmental Biology	480	339	139	2	29.1%	12.9
Cell Biology	330	229	100	1	30.4%	19.3
Women Engineers	33	3	30	0	90.9%	
Women Scientists	48	2	46	0	95.8%	

(*Source*: [9]: Appendix, A3, adapted by the authors.)

When the two predominantly female societies (Society of Japanese Women Scientists (SJWS) and the Japanese Women Engineers Forum (JWEF)) are excluded, the highest percentages of female respondents were found in the Japan Society for Cell Biology (30.4 per cent), the Japanese Society of Developmental Biologists (29.1 per cent), and the Japanese Society of Plant Physiologists (29.1 per cent). These are all biological science associations. Female respondents accounted for 15.2, 10.1 and 8.0 per cent in the Chemical Society of Japan, the JPS and the JSAP, respectively. From the relationship between the ratio of female membership and the respondent rate, we can estimate that more women were eager to return the survey in every society ([9]: 5). However, the response rates of each society are broadly in line with differing female memberships across STEM domains. The overall figure also reflects the generally low female membership in STEM (see Figure 4.1).

One of the major findings from the survey was a clear gender gap in the treatment of men and women in the fields. Gender inequality in the rank of positions was found, but even greater inequality was identified in the allocation of basic resources necessary for research and development, such as the number of subordinates and funding ([9]: 47). This issue of gender gap in the allocation of research resources was more acute in universities and public research institutions, compared with that in corporations.

The report noted the fact that female researchers and engineers have far fewer children than men, which is considered 'almost abnormal' ([9]: 47). Many female

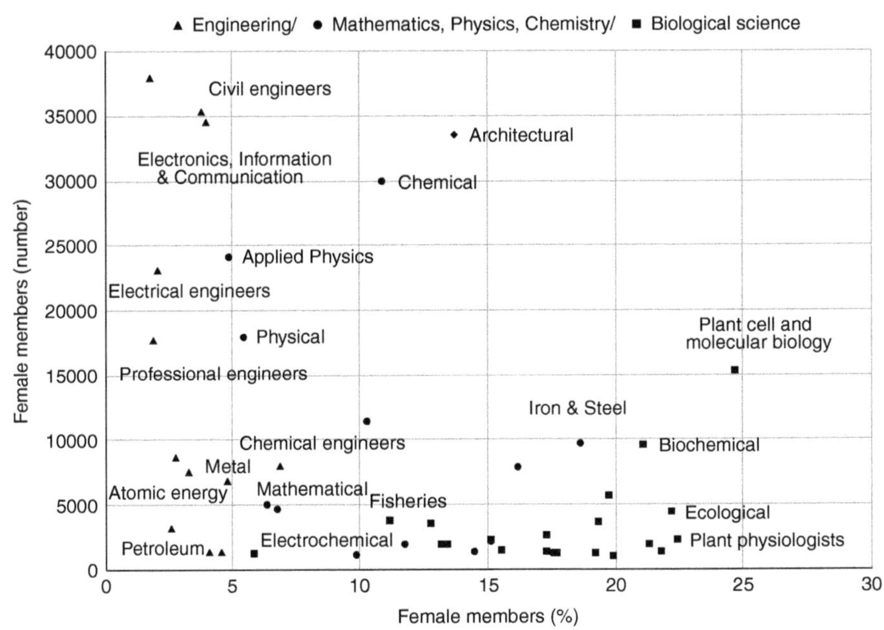

Figure 4.1 Size of STEM-related societies and their ratio of female membership.

Source: [9].

respondents selected 'change in men's mindset' and 'men should balance family and careers' as requirements for a gender-equal society. The report highlighted 'the long-standing tradition that made family affairs and childbearing the responsibility of women, and locking the roles of men and women in place'. ([9]: 47)

The SCJ responded to this initiative, and on 24 November 2004 organised a one-day symposium entitled 'How far have we come? Equal participation of men and women in academia' under the leadership of the SCJ member and the former President of the JSAP, Toshio Gotō. The event attracted more than 250 participants from various central ministries, universities and research institutions.

As previously mentioned, the Gender Equality Bureau has been publishing a White Paper annually since 2000. The White Paper, published in May 2005, focused on gender (in)equality in science and technology domains in Japan. The EPMEWSE survey report's findings and recommendations were featured in this issue.

The informed source reported that the survey report provided a strong 'evidence base' for supporting the Cabinet Office's decision to set targets. The White Paper also reported the international comparison on the percentage of female researchers. Japan ranked lowest amongst the 30 OECD member countries in terms of the percentage of women in the scientific workforce (11.1 per cent) ([10]). Austria and Germany were the only two other listed countries with the ratio of female researchers below 20 per cent (18.8 and 15.5 per cent, respectively).

The government approved the policy recommended by the White Paper in June, 2005. The source noted this cabinet approval as a 'critical moment', as there was a great deal of backlash against gender equality measures observed at that time ([11]).

Although the 2nd Science and Technology Basic Plan (March 2001) already called for expanding job opportunities for female researchers, the slow progress was critically viewed. More effort in tracking the number of successful female applicants for competitive research fund projects was deemed necessary, and this led to the recognition that the government needed to set targets to be achieved, as part of positive action. The 2000 Basic Plan for Gender Equality was revised and approved in December 2005, when the 12th section (New Areas for the Promotion of Gender Equality) was added to the old Basic Plan, where gender equality in S&T was stated as a priority.

The presence of the formerly invisible Minister for Gender Equality (in Japanese, *Shōshika. Danjo Kyōdō Sankaku Tantō Daijin* means Minister for Tackling the Low Fertility Rate & Equal Participation of Men and Women) became more visible in the public domain; political scientist and SCJ member, Kuniko Inoguchi was appointed to this post by Jun'ichiro Koizumi in October 2005, and she actively campaigned for women in STEM.[3]

In 2006, the CSTP decided to set targets for the proportion of newly available permanent positions that would go to women in the 3rd Science and Technology Basic Plan in 2006 ([12]). The overall recruitment target for female researchers in the natural sciences was 25 per cent. Specific targets varied by field: 30 per cent for health and life sciences, 30 per cent for agricultural sciences, 20 per cent for physical sciences, and 15 per cent for engineering.

As shown in Chapter 3, universities had a relatively high proportion of female researchers at the time (21.5 per cent in 2006; 24.7 per cent in 2012), compared with those figures in corporations and public research institutions. Combining the figures from universities and research institutes, female researchers in the social sciences and humanities account for almost 30.0 per cent, while those in the natural sciences account for 20.7 per cent. Therefore, the MEXT, which is responsible for education and research institutions, saw the opportunity to tackle gender inequality in STEM first by addressing issues in universities and research institutions. In order to realise the targets set by the Cabinet Office, the MEXT decided to invest 500 million JPY in ten institutional projects to encourage women in STEM over three years through a competitive selection process from 2006. The government launched 'Support for best practice for the promotion of female researchers' as part of Competitive Research Grants for the Promotion of Science. To create an environment where female researchers could fully demonstrate their abilities, the government decided to select potential model cases and provide funds, through which the model could assist female researchers in balancing their research activities and child rearing.

It was also decided that 140 million JPY would be set aside to fund 2-year fellowships for 30 scientists who had put their careers on hold to start families ([10]). The new scheme was called a Restart Postdoctoral (RPD) fellowship, and administered by the MEXT's agency, the Japan Society for the Promotion of Science (*Nihon Gakujutsu Shinkō Kai*) (JSPS).

The RPD fellowship was founded 'to facilitate the sustained participation of Japanese women in science, which it does by providing female postdoc researchers with focused assistance in making a smooth transition back into the laboratory after having suspended research activities for the purpose of childbearing and infant raising'. Prior to this, the Research Fellowships for Young Scientists Program had existed. The ratio of the number of people awarded to that of applications received was approximately the same for male and female researchers. However, the problem was that there were far fewer female applicants ([13]).

The formation and survey of the EPMEWSE opened a new era of governmental positive action in supporting female scientists.

4.4 MEXT-funded programmes: 'Support for best practice for the promotion of female researchers'

The MEXT created a programme called 'Support for best practice for the promotion of female researchers' (*Jyosei Kenkyūsha Shien Model Ikusei*), funded by one of the Special Coordination Funds for Promoting Science and Technology (hereafter referred to as the 'Model' programme). The funds were aimed at universities and research institutes which seek to build a model for promoting the equal participation of men and women in science and technology, while supporting the balance between work and family life. Each programme lasts for three years.

- Target: Researchers at universities, inter-university research institutes, national research organisations, and independent administrative agencies (natural

sciences, or inter-disciplinary areas between natural sciences and humanities/ social sciences).
- Budget: $180–450 thousand/year (including indirect expenses).

In response to the first call from the MEXT, 36 universities and research institutions submitted an application at the competitive tendering process. As a result of a two-stage selection process, the ten universities with successful bids were: Hokkaidō, Tōhoku, Tokyo University of Agriculture and Technology, Waseda, Ochanomizu, Tokyo Women's Medical University, JWU, Kyoto University, Nara Women's University, and Kumamoto University.

This policy was responsive to the findings of the EPMEWSE report, with a particular request for providing support for female scientists of childbearing age. In the report, they identified the discontinuity, and break from their research activities as the major obstacle to female researchers in science and engineering. Scientific research advances at a fast pace, and researchers often need to remain at the coalface, collecting data in a laboratory setting. The policy was innovative in that the government decided to intervene in higher education institutions, both public and private, with the objectives of prompting the establishment of a family-friendly 'research environment and changing mentality' among staff ([14]).

Leading up to this policy, the informed source mentioned that there was some persisting prejudice and hesitation towards this positive action on the part of some policymakers. Some suggested that science and technology are more suitable for men, which could explain the small number of female researchers in STEM, while others expressed anxiety that the policy may send the wrong signal which could be detrimental to the traditional division of labour between men (at work) and women (at home).

It is worth noting that these policy discussions took place when women's underrepresentation in science careers was also fiercely debated in the United States. They were triggered by the speech given by the former President of Harvard University, Professor Lawrence Summers, in January 2005 ([15] [16] [17]). In January 2007, the Minister of Health in Japan had to resign from his post after making a reference to women as 'birth-giving machines' in a public meeting. These cases demonstrate not only how social norms persist but also how flammable the issue of gender equality can be in any country. However, the large-scale survey of EPMEWSE, scientifically conducted by both men and women, gave a strong evidence base, which was undisputed ([18]).

This policy continued in its original form for five consecutive years until 2010, with ten or more institutions selected annually. At the end of the first round (2006–09), the MEXT created the 2nd phase programme called 'System for Fostering Female Researchers to Accelerate Reforms'.

> From the point of view of nurturing and ensuring diverse talents and promoting gender equality, (this programme seeks to) accelerate the process of nurturing female researchers in the disciplines (e.g. sciences, engineering and agriculture) where the ratio of female intake is low ([14]).

This project was highly ambitious, and one term was for five years (40 million JPY for the first year, and 80 million JPY from the second year onwards). Five national universities (Kyoto, Kyūshū, Tōhoku, Hokkaidō, Tokyo University of Agriculture and Technology) were selected for 2009–2013, and seven universities for 2010–2014 (Kobe, Nagoya, Chiba, Hiroshima, Kumamoto, Tokyo and Nara Women's). This programme adopted a more hands-on approach, aiming to produce immediate and short-term results. As the title of the programme ('accelerate reforms') suggests, the funds were directly used to hire more female staff. For this reason, the programme was generally directed at large- and medium-sized national universities. No private university was selected.

Since 2011, the whole programme was revamped and renewed with a new name, 'Research Activities Support Project for Female Researchers'. Two ('general' and 'hub') types were created. The former type is the continuation of

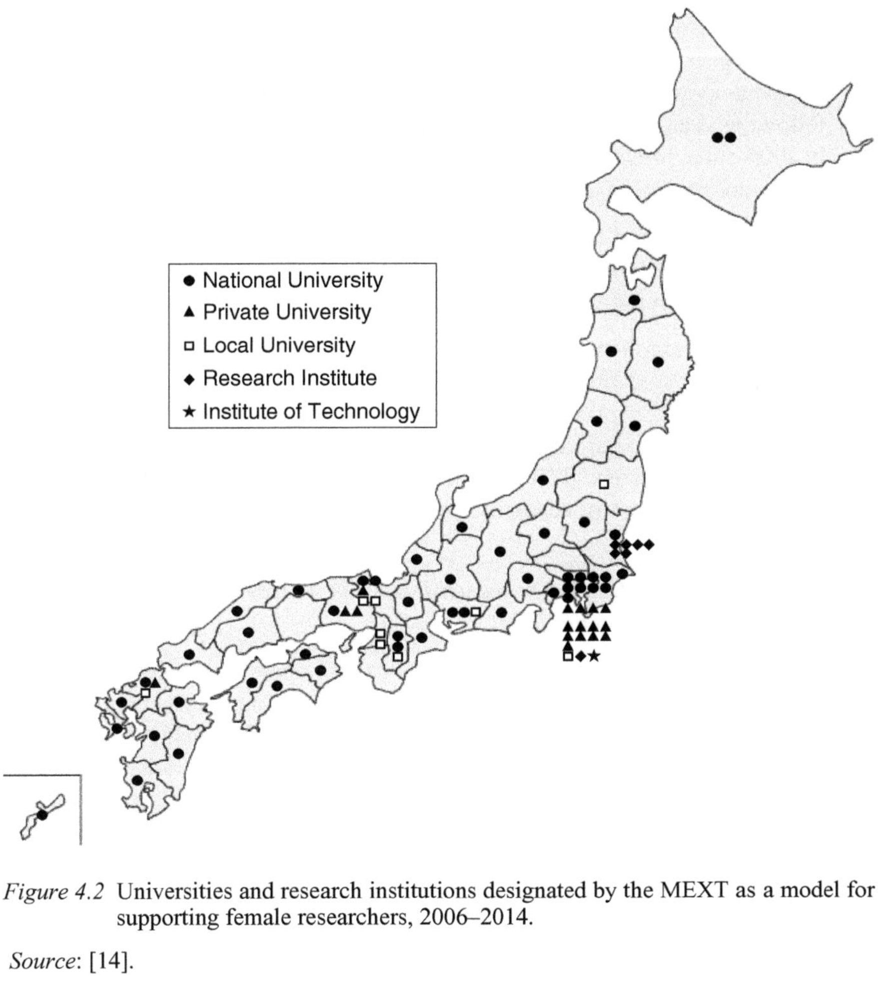

Figure 4.2 Universities and research institutions designated by the MEXT as a model for supporting female researchers, 2006–2014.

Source: [14].

the previous programme ('Support for best practice for the promotion of female researchers'), while the latter is a new model to further collaboration among multiple agencies (universities, research institutions and industries). By 2014, 95 universities and research institutions were funded through this scheme. Figure 4.2 shows the spread of these institutions.

The remainder of this section describes some unique aspects of the funded models from four selected examples (Tōhoku, Hokkaidō, Tokyo University of Agriculture and Technology, and JWU).

Tōhoku University: Science Angels

- Tōhoku University was the very first Imperial National University to admit three female students (Ume Tange, Raku Makita and Chika Kuroda) to the Faculty of Science in 1913 under the Presidency of Masatarō Sawayanagi.
- The Office for Women Researchers was established in 2001. The Declaration of Tōhoku University for Promoting Gender Equality was published in 2002. It established the Tōhoku University Gender Equality Promotion (Sawayanagi) Award among other things. The Gender Law and Policy Centre opened in the following year, while an on-campus nursery was founded in 2005.
- In 2006, three programmes were set up once the university was selected as one of the ten 'model' institutions for the 'Model' programme funded by MEXT.

  (1) Child Care and Family Care Support Programme
  (2) Research Environment Improvement Programme
  (3) Next-Generation Support Programme

As part of the Next-Generation Support Programme, the university appointed their postgraduate students as 'Science Angels'. They were formed as a group of 'role models' for high school girls, sending out the message that 'science is fun' and 'female researchers are cool'. They took part in seminars and public events to spread these messages. Tōhoku University was awarded again in 2009 for accelerating the increase in academic posts for women ([19]).

Hokkaidō University: HoMe (Research Posts For Partners of Hokkaidō University Membership)

Hokkaidō University began its support activities for female researchers in response to the BAGES. In May 2004, the mid-term goals and medium-term plan were formulated. The goals included 'increasing the proportion of female teachers' and 'equipping facilities and creating an environment for research staff and postgraduate students, postdoctoral fellows, and foreign researchers to be able to conduct research while raising their children'. A nursery called the Garden of Children was set up in 2006 to fulfill these aims.

- The Ōtsuka Award was created for excellent female doctoral students, and as part of the twenty-first-century Centre-Of-Excellence (CEO) programme, female researchers were supported in Fisheries Sciences Research Institute. But until female researchers came together across the whole university to discuss challenges and difficulties in August 2005, it was not yet a university-wide movement. In 2006, the university was awarded the MEXT 'Model' project, and the Support Office for Female Researchers in Hokkaidō University (FResHU) was founded directly under the authority of the President's Office in July 2006 ([20]).

- On top of career and counselling services, the Support Office in Hokkaidō is engaged in specific activities and support measures for female scientists in two areas. Therefore, a variety of support measures are classified in these two areas: 'concrete measures for increasing the number of female researchers' and 'environment creation for female researchers'. A difficulty common to both areas is that of starting a family.
- In dual-career marriages, the ability to find employment for both couples is a major challenge. If either a woman or her partner cannot find a research post in the same or neighbouring area, one of them must go alone, or the female researcher often gives up her career. They found that only 10 per cent of male researchers have a partner in research, while 50 per cent of female researchers have a partner in research. They are faced with the difficult choice of giving up a career or living apart.
- Since Hokkaidō is a remote region with few job opportunities for researchers, apart from its capital Sapporo, the Support Office provides a research post for a researcher's partner as well as support during their parental leave. This scheme is called HoMe (Research Posts for Partners of Hokkaidō University Membership).

Tokyo University of Agriculture and Technology (TUAT): One-Plus-One

TUAT was also one of the first-wave universities funded by the MEXT grant in 2006. In 2009, the university was awarded the 'Acceleration' programme ('System for Fostering Female Researchers'), along with four other large former Imperial Universities (Kyoto, Kyūshu, Tōhoku and Hokkaidō). TUAT, located in the western part of Tokyo, has only two faculties, Agriculture and Engineering, and female researchers are seriously underrepresented in both.

In 2009, the TUAT undertook a unique positive action called One Plus One in order to actively increase female researchers. This system works as follows. When a department appoints a female researcher, lecturer, associate professor or professor, the university provides money to hire another staff member. Three conditions must be met.

- The 'Plus One' staff member does not have to be a woman, and (s)he will be affiliated with the department who made this appointment.
- The 'Plus One' staff member does not have to work with the appointed female staff member.

- As the scheme encourages the appointment of a woman in a higher position, the job classification of the 'Plus One' staff's position would depend on her level.

Japan Women's University: U-Researchers

As JWU is the only private women's university with a faculty of science, it was selected as one of the ten first-wave institutions by the MEXT. The project was named 'The Multi Career Path Support Model for Female Researchers', and had three streams: 'Support for Ubiquitous Research (U-research)', 'Support for Human Resources' and 'Research and Planning'. The three streams were designed to support a better balance between research and family life.

Making full use of IT technologies, such as computer networks and high-speed broadband, it helps women conduct research activities wherever they are at any time. By introducing a system of part-time research assistant and setting up a video conferencing system, 'ubiquitous' researchers (U-researchers) were hired to continue their research from home, while looking after their children. As an example, one graduate from JWU with a Master's in Science was hired as a U-researcher. After graduation, she worked in the R&D division of an electronics company, but made a decision to leave the job when she became pregnant.

These MEXT-funded models yielded both tangible and intangible outcomes, and in 2012, the MEXT carried out evaluation, which will be the focus of the next chapter.

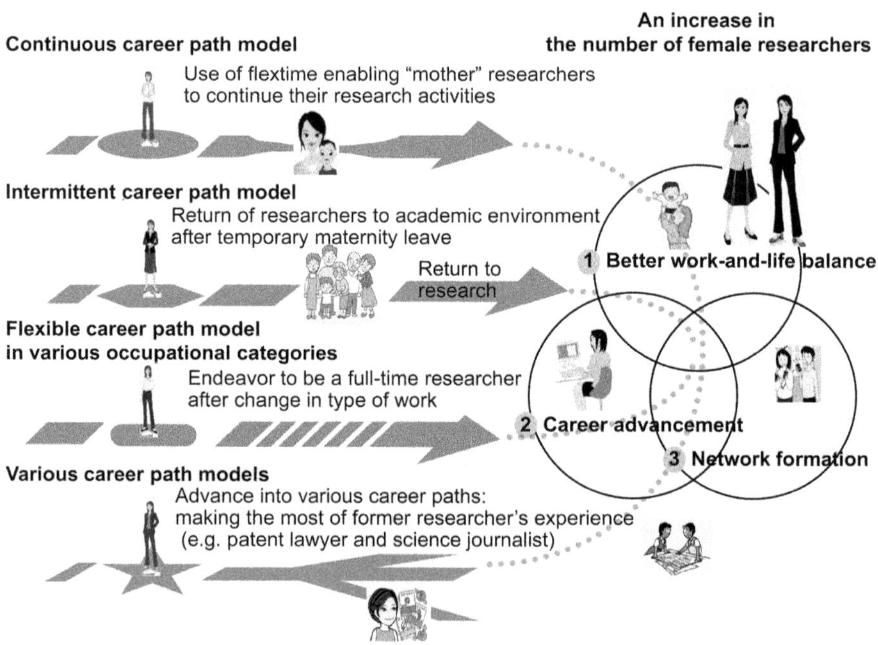

Figure 4.3 Multi-career paths, envisaged by JWU.

4.5 Other types of support for women in STEM

Other policy initiatives were taken by the government and its agencies.

1. *Super Science High School.* In 2002, MEXT launched a new accreditation system called Super Science High School (SSHS), and it annually designates several High Schools which prioritise science and mathematics. In the first year, 26 out of 77 schools were selected. Initially, one term was decided to be three years, but since 2005, it became five years. In 2002, 700 million JPY was allocated for this project, but the amount has since increased to 2.4 billion JPY in the financial year 2011.
2. *Summer School for Junior and High School Girls.* Another Independent Administrative Institution, answerable to the MEXT, is the National Women's Education Centre (NWEC). NWEC started a summer school, targeting junior- and high-school girls from 2006.

    - *Target*: 3rd year junior high school girls; senior high school girls; and parents and teachers with an interest in science and technology.
    - *Content*: a residential hands-on science programme with the aim of enabling junior and senior high school girls to discover new possibilities for themselves in the fields of science and technology through dialogue and interaction with scientists, engineers and students at science and engineering universities. The school also provides a programme for parents and teachers in conjunction with this.

3. *Restart Postdoctoral (PRD) Fellowship.* The RPD Fellowship, managed by the MEXT's agency, the JSPS, was mentioned earlier. The RPD fellowship is annually awarded to about 50 post-doctoral researchers. Now female researchers in social science as well as natural scientists are eligible for this fellowship.

    - *Target*: All fields of the humanities, social sciences and natural sciences.
    - *Number of new awardees per year*: About 50.
    - *Eligibility*: Postdoctoral researchers who have within the past 5 years suspended their research for a period of 3 months or longer for the purpose of child birth and infant nursing. No restriction on age or gender.
    - *Tenure*: 3 years.
    - *Monthly stipend*: 362,000 JPY.

4. *Funding Programme for Next Generation World-Leading Researchers (NEXT Programme).* In 2009, another funding stream was created by the JSPS, based on the government's 2009 financial year supplemental budget. 'Aimed at advancing cutting-edge R&D of great significance for policy and society, the programme provides a system of research support for young researchers with expected potential to become world leaders in their respective fields of science and technology' ([13]: 15). Grants are awarded to young and female researchers (25 per cent of all grants). It was the first time positive action was employed for such a competitive research grant in order to meet the target of 30 per cent set by CSTP.

- Eighty-six per cent of applicants (4,818) are university researchers, while 52 (1 per cent) are from a private companies, and 637 (11 per cent) are from public research institutions. More than 1,500 female researchers applied (1,588 in total), 82 of whom were awarded grants (25 per cent) (see Table 4.2).

The Japan Science and Technology Agency (JST) is an independent administrative institution (IAI) that is held accountable to the MEXT, and plays a key part in deciding who is awarded competitive research funding.

5. *Promoting Awareness of Female Researches.* The Office for Promoting Equal Participation of Men and Women was established within JST in April 2006. The Office promotes awareness among female researchers and encourages them to apply for research funding. It also publishes a booklet featuring various female researchers and practitioners in science, engineering and technology, entitled *Role Models: A bright future for women in science, engineering and technology.*
6. *'Childbirth, Child Care and Home Care' Support Programme.* The funding is awarded to researchers who are involved in JST's strategic research

Table 4.2 Number of NEXT Programme grant applicants and successful candidates by gender

	Green Innovation		*Life Innovation*		*Total*	
	Awarded	*Applicants*	*Awarded*	*Applicants*	*Awarded*	*Applicants*
Male	110	1,893	137	2,167	247	4,060
Female	31 (22%)	505 (21%)	51 (27%)	1,053 (33%)	82 (25%)	1,558 (28%)
Total	**141**	**2,398**	**188**	**3,220**	**329**	**5,618**

(*Source*: [13].)

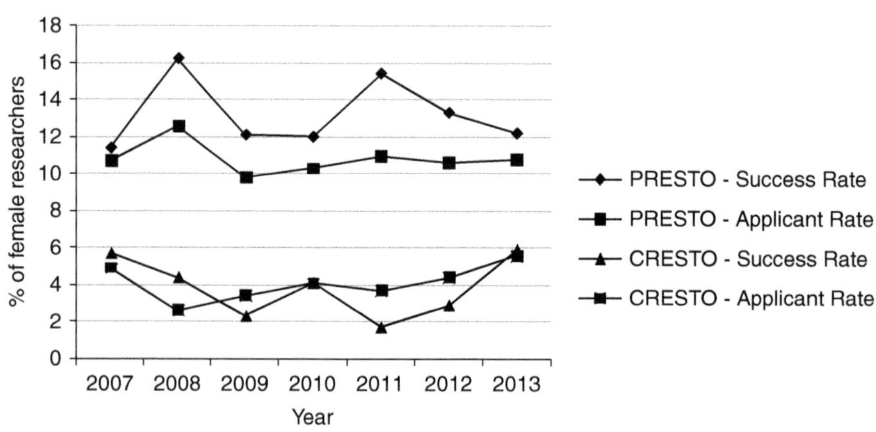

Figure 4.4 Ratio of female applicants and success rate in obtaining two types of JST funding.

Source: [14].

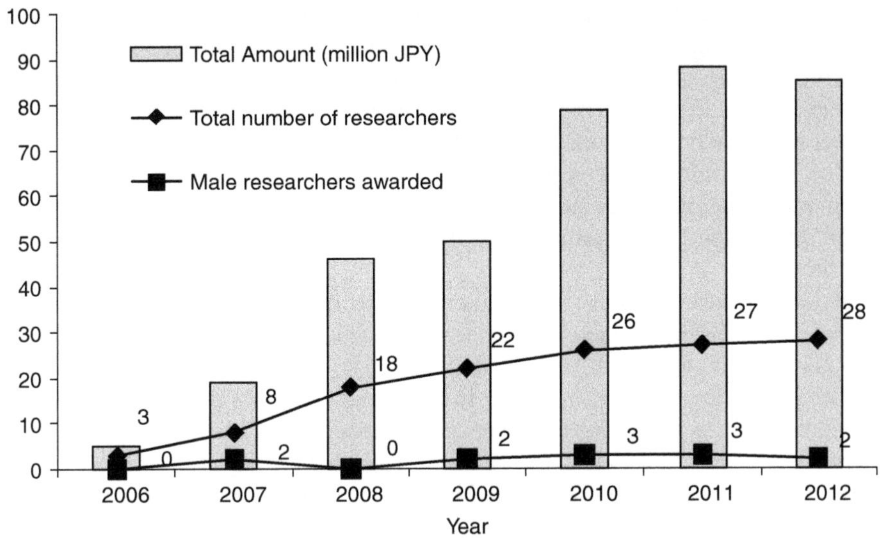

Figure 4.5 Number of researchers awarded and total budget spent (2006–2012).

Source: [14].

programmes such as CREST, ICORP and SORST. The eligible applicants are at various stages of life and faced with life events (childbirth, child care and care for the elderly). With this support, they will be able to continue their career.

- CREST: Core Research for Evolutional Science and Technology
- ICORP: International Cooperative Research Project
- SORST: Solution-Oriented Research for Science and Technology

7. *'Encouraging Secondary School Girls to Specialise in Sciences' Support Programme*. The funding will be given to organisers to plan and host an event for girls at secondary-level education, with the aim of stimulating their interest in undertaking science research at a higher level of education. This is funded by the Cabinet Office.

4.6 Activities of scientific societies: Japan Society of Applied Physics

As previously mentioned, the JSAP was the very first academic society to establish the Committee for Promoting Equal Participation of Men and Women in Science and Technology. Since 2001, the JSAP has held spring and autumn academic lectures (twice every year) on the themes of human capital resources and equal participation in science and engineering. In April 2006, two sub-groups (students and young researchers, and seniors) were added to the Committee for Promoting Equal Participation of Men and Women in Science and Technology, which was renamed the 'Committee for Promoting Equal Participation of Men and Women and Investing in People'.

The committee aims to promote diversity in the fields of science and technology by providing everyone, regardless of gender and age, with opportunities for making the most of their talents. The committee consists of three sub-committees: (i) equal participation of men and women; (ii) supporting students and early-career researchers; (iii) contributions to wider society.

The newly reformed committee embarked on various projects. They have been providing nurseries at spring and autumn academic conferences since 2005. More than 100 conference attendees (working parents) used the facility in the first year.

JSAP's working group on Industry-Academia-Government Collaboration launched a website 'Women in Science and Technology' (featuring role models), and is now part of the JSAP official website [22].

Career consultations for graduate students and post-docs provide a chance to speak directly to recruiters from various companies and research institutes for graduate students and post-docs. These services have been offered twice every year at spring and autumn academic lectures since 2008. The JSAP also created a 'career explorer' mark, which postdoctoral researchers can attach to their presentations at these conferences, and signal their intention of looking for a job.

The JSAP also designed a roadmap called 'Vision for the Scientists of Tomorrow' in 2008. The roadmap has four specific purposes: to expand opportunities for scientists of future generations; to increase the number of female scientists; to promote scientists' status in different sectors, and to enhance collaboration between industry and academia. It was revised in 2010. It also promotes science amongst junior and high school students at its spring and autumn conferences. In 2010, special lectures, a 'Science Adventures' experiment corner and poster sessions were offered, engaging secondary school students through interactions with senior and young researchers.

In 2007, to mark the JSAP's 75th anniversary, it held at 'Science and Technology Exhibition: applied physics that changes the world' on 3 and 4 August. This was targeted at both adults and children, and attracted more than 8,000 people.

EPMEWSE is also very active in promoting gender equality. As of 2013, full membership had been given to 52 academic societies, and observer status to 28 societies. They have carried out their second and third large-scale surveys in 2007 and 2013, respectively.

Table 4.3 summarises various activities that were initiated in the early 2000s by government ministries and other advisory bodies, in order to promote gender equality in STEM, and particularly to achieve its target numbers.

4.7 Conclusion

This chapter has covered a period with a myriad of policy initiatives and changes. Global competition among higher education institutions intensified, particularly after the 2000s, with the publication of various benchmarking initiatives and rankings. In particular, the GEM highlights the equal opportunities in political and economic activities for women (e.g. the proportion of female legislators,

Table 4.3 Government and other stakeholders' activities in promoting gender equality in STEM

Cabinet Office	*MEXT*	*Council for Gender Equality (Expert Committee, Cabinet Office)*
December 2000: 1st Basic Plan for Gender Equality (Cabinet Approval) January 2001: the Gender Equality Bureau established. Minister appointed. May 2005: White Paper on Equal Participation of Men and Women (female underrepresentation in STEM highlighted) December 2005: 2nd Basic Plan for Gender Equality (Cabinet Approval) 2005: Website portal open to support women who take the challenge ('Challenge Campaign')	March 2003: 'Women's diverse careers that change society' (1st Report) published. First mention of '30 per cent' as the target ratio of female researchers working in universities and research institutions. April 2003: Forum to discuss 'a variety of career options' for women	October 2002: under the leadership of the then Prime Minister J. Koizumi, various policies to support women who take the challenge ('Challenge Campaign') were discussed. April 2003: Recommendations for creating a fairer and more transparent human resources system, particularly in recruiting and evaluating female researchers' performance.
Japan Association of National Universities	*Science Council of Japan*	*EPMEWSE*
May 2000: Recommendations to promote gender equality in the national universities adopted. Recommendations included positive action to achieve the goal of raising the ratio of female staff members to 20% in the national universities by 2010	June 2000: At the 132nd General Assembly, it adopted a statement regarding specific measures to improve the research environment for women, and published a statement 'Promotion of gender equality in the SCJ'. Special Committee on Gender issues established. October 2004: Symposium, 'How far have we come with gender equality?'	October 2002: the Japan Inter-Society Liaison Association Committee for Promoting Equal Participation of Men and Women in Science and Engineering established. 2003: First large-scale questionnaire survey (41 academic societies, respondents 19,291) 2007: Second large-scale questionnaire survey (65 academic societies, respondents 27,318)

(*Sources*: Cabinet Office, EPMEWSE, JANU, MEXT, SCJ websites. Adapted by the authors.)

senior officials, and managers, and women in professional and technical positions in each country). Moreover, the relationship between gender inequalities and economic growth began to be widely researched.

National governments worldwide have been responding to this global trend and, with varying degrees of enthusiasm, adopting policies to promote the status of women and gender equality. Japan has been no exception. Yet in STEM fields, the under-representation of female scientists was so acute that there was no guarantee that the government would take heed of the voice of the minority.

Social consensus and academia's initiatives remain two indispensable factors for the initiation of successful policy changes by any government. The government's agenda of equal participation of men and women finally became aligned with academic societies' joined-up efforts to reverse the low ratio of female scientists in their fields, when the EPMEWSE was formed in 2002.

In many respects, the underrepresentation of women in STEM has the same underlying cause as many other social issues, including the Japanese traditional view of gender roles. Under the banner of 'equal participation of men and women', the EPMEWSE's large-scale questionnaire provided a fair and objective picture of how both men and women in STEM struggle to strike a good balance between work and family life.

As a result of all three of these elements (international pressure, social awareness and academic initiatives), a breakthrough in policymaking ensued, and positive actions towards women in STEM have been implemented since 2006. And yet, there is a long way to go befor meeting the quantitative targets.

The 3rd Science and Technology Basic Plan in 2006 ([12]) set the overall recruitment target for female researchers in the natural sciences at 25 per cent. Specific targets varied by field (30 per cent for health sciences, 30 per cent for agriculture, 20 per cent for sciences, and 15 per cent for engineering). In 2009, in the university sector, the actual figures are 34.6 per cent for health and life sciences, 16.3 per cent for agriculture, 12.7 for sciences, and only 5.9 per cent for engineering. ([21])

The next chapter will ask the question: what impact have these initiatives had on female scientists, and institutions, looking at various reports and questionnaire results.

Notes

1 They are Fumiko Yonezawa (physicist/former and first female President of the Physical Society of Japan) and Masako Niwa (life scientist/former President of Nara Women's University).

2 Bandō is an alumna of Kyoto University. She later became the second female President of the JPS between September 2006 and August 2007. The first female President was Professor Emeritus of Keiō University, Fumiko Yonezawa (1996–1997) who was also an alumna of Kyoto University.

3 It is ironic that the original ministerial post for Gender Equality on its own was abolished and merged with that for Tackling the Low Fertility Rate (created in 2003).

References

[1] S. Takahashi. *Japanese Responses to Globalization: Politics, Security, Economics and Business*, chapter 'The impact of globalization on domestic politics in Japan', pages 35–54. Basingstoke: Palgrave, 2006. eds. Glenn Hook and Harukiyo Hasegawa.

[2] H. Takenaka. *Kōzō Kaikaku no Shinjitsu: Takenaka Heizō Daijin Nisshi (The Truth about the Structural Reforms: The Ministerial Diaries of Takenaka Heizō)*. [In Japanese] Tokyo: Nihon Keizai Shinbunsha, 2006.

[3] Y. Uchiyama. *Koizumi Seiken: 'Pathos no Shushō' ha Nani wo Kaetanoka? (Koizumi Administration: What has 'the Prime Minister of Pathos' changed?)*. [In Japanese] Chukō Shinsho. Tokyo: Chūō Kōronsha Publisher, 2007.

[4] K. Takayasu. Prime-ministerial power in Japan: a re-examination. *Japan Forum*, 17(2): 163–184, 2005.

[5] K. Nakano. Cross-national transfer of policy ideas: Agencification in Britain and Japan. *Governance*, 17(2): 169–188, 2004.

[6] N. Kodate, K. Kodate, and T. Kodate. Mission completed? Changing visibility of women's colleges in England and Japan and their roles in promoting gender equality in science. *Minerva*, 48(3): 309–330, 2010.

[7] Interview with former President of Ochanomizu. Personal communication.

[8] EPMEWSE. [In Japanese]. [online]. Available from: http://www.djrenrakukai.org/organizations.html. Accessed 13 December 2014.

[9] EPMEWSE. Diverse visions of scientists and engineers in the 21st century: For the promotion of gender equality. Technical report, Report on the MEXT-commissioned project, 2003, 2004.

[10] National Institute of Science and Technology, MEXT. Kyōiku Shihyō no Kokusai Hikaku (International Comparison of Education Standard) [In Japanese]. Technical report, Tokyo: MEXT, 2005.

[11] Interview, former Director, Gender Equality Office, JST. Personal communication.

[12] Cabinet Office. Dai san-ki kagaku gijyutsu kihon keikaku (The 3rd Science and Technology Basic Plan) [In Japanese]. [online]. Available from: http://www.cao.go.jp/cstp/english/basic/index.html, 2006. Accessed 10 June 2008.

[13] Japan Society for the Promotion of Science. Support for female researchers in Japan. [online]. Available from: http://www.jsps.go.jp/english/e-quart/18/, 2012. Accessed 13 December 2014.

[14] Japan Science and Technology Agency. Kagaku gijyutsu ni shisuru jinzai ikusei ni kansuru shisutemu kaikaku (Systemic Reform for Human Resources Development for Science and Technology) [In Japanese]. [online]. Available from: http://www.jst.go.jp/shincho/program/woman_ken.html, (n.d.). Accessed 15 November 2014.

[15] S. Goldenberg. Why women are poor at science, by Harvard president. *The Guardian*, January 18, 2005. [online]. Available from: http://www.theguardian.com/science/2005/jan/18/educationsgendergap.genderissues. Accessed 27 November 2014.

[16] N. Hopkins. Diversification of a university faculty: Observations on hiring women faculty in the schools of science and engineering at MIT. MIT Faculty Newsletter, (March/April 2006). [online]. Available from: http://web.mit.edu/fnl/volume/184/hopkins.html. Accessed 27 November 2014.

[17] M. Mōri, K. Higuchi, S. Aoyama, H. Kunii, K. Chiba, K. Murofushi, and Ochanomizu Academic Association. Jyosei to kagaku wo kagaku suru (Scientific Analysis of Women in Science) [In Japanese]. Tokyo: Ochanomizu Academic Association, 2006.

[18] Interview, former Director, Gender Equality Bureau, Cabinet Office. Personal communication.

[19] Tōhoku University Centre for Gender Equality Promotion. [online]. Available from: http://www.morihime.tohoku.ac.jp/about/torikumi/, (n.d.). Accessed 13 December 2014.

[20] Support Office for Female Researchers in Hokkaido University. Katsudō naiyō (Programmes and activities) [In Japanese]. Available from: http://freshu.ist.hokudai.ac.jp/activity/. Accessed 13 December 2014.

[21] National Institute of Science and MEXT Technology Policy. Kagaku gijyutsu wo meguru syuyō kakkoku nado no seisaku dōkō bunseki (Trend analysis of science and technology policy in major countries)[In Japanese]. Technical Report 117, NISTEP Report, 2009. [online]. Available from: http://www.nistep.go.jp/achiev/ftx/jpn/rep117j/idx117j.html. Accessed 22 November 2014.

[22] Japan Society of Applied Physics. Rikei shusshin no jyosei tachi (Women in science and engineering) [In Japanese]. Available from: http://www.jsap.or.jp/gender/careerpath/, (n.d.). Accessed 23 April 2015.

5 Policy impacts and voices of women in science and engineering

Evaluation and survey results

5.1 Introduction

The previous chapter highlighted a number of new policies and initiatives that were introduced in the mid-2000s. This chapter will look at the impact of those measures on female scientists and institutions by reviewing some key reports and documents. The MEXT-funded 'Model' programme (introduced in 2006) has been the most encompassing policy to date in Japan, covering all types (national, local and private) of university and national research institute across the country. Since 2013, firms and government bodies can also join the programme as partner institutions.

In order to monitor the progress in gender equality in the university sector, the SCJ conducted three large-scale surveys on all universities in 2007, 2010 and 2013. The first section of this chapter reviews the results and findings from these surveys. As the MEXT programmes target only the higher education and research sectors, excluding industry, the data will be complemented by the survey findings conducted by EPMEWSE (the Japan Inter-Society Liaison Association Committee for Promoting Equal Participation of Men and Women in Science and Engineering).

The second section of the chapter will present the results of recent questionnaires submitted by the alumnae who specialised in mathematics and science at two women's universities. The questionnaires reveal what graduates think about themselves (women studying science), and how they reflect upon their own learning and its impact on their life and career after graduation. Both the benefits and challenges of female science students are highlighted. The chapter also features some women who have had research experience overseas. The aim of the chapter is to investigate progress not only in the 'why so few' question, but also in more qualitative aspects of individuals' lives.

5.2 Impact and changes

5.2.1 Science Council of Japan's large-scale survey results

SCJ has conducted a large-scale survey every three years on all the universities in Japan (2007, 2010 and 2013). The survey carried out in 2007 was the first one after the introduction of the MEXT-funded programmes. It helped to shed light on the work environment and policies of each institution with regard to supporting female researchers and academics (see Table 5.1).

The total number of universities in each of the three categories (contacted for the SCJ surveys) is shown in Table 5.2.

The first survey results were published in late 2007. The survey has historical significance in the Japanese higher education sector, as it was the first time that all three types of university were included in the same survey concerning gender equality. Previously, surveys were conducted separately by organisations such as the National Universities Association (Kokuritsu Daigaku Kyōkai, Kokudaikyō) and the Association of Private Universities of Japan (Nihon Shiritsu Daigaku Kyōkai).

The data suggest that a much higher proportion of national universities responded to all three surveys, although the response rates dropped sharply the third time. This drop occurred partially because the National Universities Association conducted a separate survey regarding gender equality in 2013.

The findings from the first survey revealed that while many universities had established committees, policies and guidelines to deal with sexual harassment and family care leave, few universities proactively sought information to establish the challenges facing female researchers in their own institutions. In particular, only 0.6 per cent of private universities (which responded to the survey) had collected such data internally at the time of the first survey ([1]).

When the first survey was conducted, 20 universities and institutions across the country were selected by the MEXT as 'model' institutions. The differences between those selected for this programme (n=20), and the rest (n=433) turned out to be remarkable. Seventy-five per cent of the 'model' institutions promoted

Table 5.1 SCJ large-scale surveys

	Survey period	*Total no. of universities contacted*	*No. of universities that responded*	*Response rate (%)*
1	June–July 2007	705	453	64.3
2	May–June 2010	733	460	62.8
3	Aug.–Oct. 2013	775	443	57.2

(*Sources*: [1], [2], [3]. Adapted by the authors.)

Table 5.2 Number of universities contacted for the SCJ survey: by university type (2007, 2010 and 2013)

	2007	*2010*	*2013*
Private universities	544 (77%)	570 (77%)	606 (78%)
Local universities	75 (11%)	77 (11%)	83 (11%)
National universities	86 (12%)	86 (12%)	86 (11%)
Total	**705**	**733**	**775**

(*Sources*: [1], [2], [3]. Adapted by the authors.)

Table 5.3 SCJ large-scale survey response rates by university type (2007, 2010 and 2013)

	2007	*2010*	*2013*
Private universities	57.9%	56.1%	56.4%
Local universities	78.2%	72.7%	48.1%
National universities	89.5%	99.7%	70.9%

(*Sources*: [1], [2], [3]. Adapted by the authors.)

gender equality, and disseminated information regarding the issue, while only 16.9 per cent of the non-selected institutions engaged in such activities. Other aspects showing stark differences between the selected and non-selected institutions include: the establishment of gender equality policies and guidelines (62.5 per cent, 9.9 per cent), and the existence of boards and committees to deal with matters related to gender equality (87.5 per cent, 12.2 per cent). Participation rates in campaigns promoting science and engineering among secondary school students were much higher in the 'model' institutions (37.5 per cent) than the other group (2.5 per cent).

Out of 705 universities, the size and location vary greatly. It was also found that although some small- and medium-size universities in rural areas sought to promote gender equality, the lack of resources and funding prevented them from implementing such measures.

Furthermore, the findings demonstrated that the level of engagement among private universities in gender equality is weaker. However, since private universities had a much higher proportion of female faculty members (particularly in arts and humanities), the issue was not seen as a priority. The proportion of female faculty members in national universities was much smaller than the other two types. When it comes to positions, however, women accounted for 51.4 per cent in research assistant positions, while they accounted for only 11.6 per cent in professorial posts, as of 2008.

As indicated in the previous chapter, progress was particularly slow in STEM fields. In 2006, the government's 3rd Science and Technology Basic Plan set out the overall recruitment target at 25 per cent for female researchers in the natural sciences. In 2007, 24.3 per cent was achieved in the university sector as a whole, although the actual figures of 34.6 per cent for health and life sciences masked much lower ratios in other natural sciences (12.7 per cent) and engineering (5.7 per cent).

One indicator of the difficulties surrounding female researchers in STEM fields is the acquisition rates of competitive research funding. The success rate tends to decrease as the age of the female applicant increases. In addition, the volume of funding (5 per cent of the entire budget) is also much smaller than the proportion of successful female researchers (12 per cent). Female researchers tend to lead projects with smaller budgets.

Based on the findings from the first survey, policy recommendations 'Promoting equal participation of men and women in academic fields' were put forward in 2008 ([1]). Furthermore, in 2010, the SCJ included these recommendations in a report entitled 'Prospects of Japan–Recommendations from academia 2010'([2]).

The recommendations were approved by the General Assembly of the SCJ on 5th April 2010, and submitted to the government. In Chapter 4 of the report, the following four recommendations were outlined with regard to government policies necessary to promote gender equality in academic disciplines.

1. Legislation and administrative support: inter-ministerial collaboration, legislation and budgetary measures to improve working conditions (e.g. parental leave criteria for fixed-term employees).
2. Institutionalisation of gender equality initiatives: promotion of gender equality policies, targeted at private universities in particular, is necessary, in order to fill the gaps in policy implementation between public (national/local) and private universities, which were highlighted by the SCJ survey.
3. Continuation and further acceleration of the MEXT-funded programmes: the effectiveness of these programmes was demonstrated by higher rates of gender equality policy implementation among the universities selected as 'models' for these programmes. These efforts should be maintained.
4. Inclusion of gender equality achievement levels as part of university assessment criteria: for example, stating the gender ratio among faculty staff in their institution-based application for a research grant can incentivise the faculty to promote gender equality ([2]).

Simultaneously, the Advisory Committee for the Gender Equality Bureau, Cabinet Office had been tasked with a progress review of the 2nd Basic Plan for Gender Equality for the government. The interim report containing a set of analyses was published on 5th April 2010. The 3rd Basic Plan for Gender Equality was discussed and approved later that year in December.

They identified five main reasons for the lack of progress in gender equality in STEM fields.

1. Persistent gender stereotypes in society and few role models, discouraging girls from specialising in STEM subjects.
2. Insufficient progress made in hiring practices and promotions in research institutes and academia.
3. Insufficient investment in infrastructure for a better work–life balance.
4. Long working hours (e.g. experiments in a laboratory) prevent women from choosing STEM as a career.
5. Insufficient engagement of employers in positive action towards female employees, both in recruitment and promotion.

As mentioned in the previous chapter, the four-year 'Acceleration' programme ('System for Fostering Female Researchers to Accelerate Reforms') was rolled out in 2009 by MEXT as the next stage of the 'Model' Programme. Five national universities (Kyoto, Kyūshū, Tōhoku, Hokakidō, Tokyo University of Agriculture and Technology) were selected for the first year, followed by seven (Kōbe, Nagoya, Chiba, Hiroshima, Kumamoto, Tokyo and Nara Women's) in the second year. Between 2007 and 2010, the ratio of female researchers increased from 12.4 per cent to 13.6 per cent.

The 2nd SCJ survey was conducted in 2010, and another report entitled 'Towards the acceleration of the promotion of gender equality in academic fields' was published the following year ([3]).

Some of the results from the latest SCJ survey, carried out in 2013, will be shown below.

This time, the National Universities Association conducted a separate survey, and therefore, this survey is missing some data relating to national universities. However, the results indicate an overall increasing trend of female full-time staff across the three types of universities, except a drop between 2004 and 2007 in local universities.

The degree of commitment to promoting gender equality varies greatly across the three types of university. It was found that the two measures ('care leave system' and 'shorter working hours') are implemented widely across all the three types of university (80 per cent on average). However, implementation of a 'nursery at work-place' measure demonstrates remarkable gaps (50 per cent in national universities, while only 5 and 6.7 per cent in local and private universities have these facilities).

Table 5.4 shows other aspects of university governance relating to gender equality.

In particular, the gaps between national and private universities are stark in the areas of recruitment, quotas and governance structure (e.g. policies and committees).

In STEM fields, less than 10 per cent of respondents from private universities are promoting gender equality, and more than 30 per cent in both local and private universities answered that there is no gender inequality in STEM or other fields.

Table 5.4 Percentage of universities that implemented gender equality measures in terms of governance (by university type)

	National (%)	*Local (%)*	*Private (%)*
Setting target quota for female staff	68.3	15	2
Gender equality policies and guidelines	90	18	8
Collecting data regarding gender equality	56.7	15	9
Gender equality boards and committees	95	20	10
Publicity and promotion of gender equality	88.3	23	16
Meeting times restricted to working hours	56.7	45	38
Established public advertisement system	80	75	54
Promoting female staff to decision-making positions	70	58	56
Existence of family care leave system	96.7	100	93
Measures of sexual harassment prevention	100	100	96

(*Sources*: [1], [2], [3]. Adapted by the authors.)

Interestingly, 80 per cent or more in local and private universities find no obstacle to promoting gender equality, whereas half of the national universities recognise obstacles. This is partly attributed to the fact that more pressure is exerted on national universities to engage in gender equality measures than on private and local universities.

The types of obstacle include: the difficulty of obtaining support from administrative and academic staff, students or/and researchers; insufficient information about the programmes; and difficulty in meeting the criteria set by MEXT and other government agencies. Rather than the lack of financial or human resources, half of the national universities run into difficulties in shoring up support from students and staff, and meeting the criteria for applying for MEXT programmes. Finally, differences were also found among the selected ('model') universities, non-selected national universities and private universities in their engagement levels in the promotion of female staff.

Although national universities are generally more proactive than the other two types, there is a great difference between the selected 'model' universities and the rest of the national universities. On average, the 'model' universities (45 national and 12 non-national) are doing more than the national average in two aspects (i.e. providing support to their female staff and promoting female staff to leadership roles). Eighty per cent of non-selected national universities have not implemented any special measures in order to promote gender equality. This absence of measures underscores the impacts that MEXT programmes had on the selected universities.

Finally, the ratios of female presidents, vice-presidents and deans for the three types of universities also differ (between 2010 and 2013). Except for local universities, the upward trend is confirmed, particularly among vice-presidents and deans in national universities.

5.2.2 Evaluation of the MEXT-funded programmes

The 3rd Basic Plan for Gender Equality was approved by the Cabinet in December 2010, followed by the 4th Science and Technology Basic Plan (approved in August 2011). The 4th Science and Technology Basic Plan renewed the target ratios of female researchers in each STEM subject (20 per cent for science, 15 per cent for engineering and 30 per cent for agriculture). The 3rd Basic Plan for Gender Equality reiterated the goal (set out in 2003 by the Cabinet Office) that 'by 2020, 30 per cent of leading positions in society should be occupied by women' (so-called '202030 campaign'). The 4th Science and Technology Basic Plan also aligned its goal with this 30 per cent target, and stated that the average proportion of women in STEM across different disciplines should be 30 per cent ([4]: 35). The Basic Plan also outlined the following policy initiatives for the future.

In order to facilitate the continuation of a research career, a database was to be created and provided for free to enable job search in collaboration with industry, academia and government. Furthermore, a support mechanism should be provided for female researchers who returned to work after childbirth and during the child-rearing period. For example, flexible working hours, a fair but mother-friendly promotion system, and personnel support, including the temporary hiring of research or auxiliary staff, should be established at universities and public research institutes.

Positive action with a clear target and information disseminated regarding what has been achieved so far were also mentioned as required measures.

In addition, the issue of recruiting talented female students into STEM fields was addressed. The importance of role models, the enhancement of science facilities and infrastructure in schools, and appropriate teaching methods to invigorate students' interest in science and technology were among those issues highlighted. The plan emphasised the government's commitment to programmes such as a SSHS. ([4]: 36)

In 2012, MEXT published an evaluation report on 'Special Coordination Funds for Promoting Science and Technology' ([15]). This report contains MEXT's own review of the 'Model' programme, for which 55 institutions were awarded (universities and public research institutes).

The report states five goals of the programme: (a) establishment of a support mechanism, (b) development of a research environment, (c) raising awareness and promotion of gender equality, (d) positive action, and (e) measures to support next-generation researchers ([15]: 13). It also examined the impact of the policy, mainly in the following three major areas:

1. Continued efforts and retention levels (1.1) Sustaining research activity levels among female researchers during maternity and the childrearing period (1.2) Retention of female researchers in the course of maternity and the childrearing period.
2. Increase of female staff members in the institution.
3. Increase of female students at undergraduate and postgraduate levels in STEM fields.

(1.1) and (1.2) broadly correspond to (a) and (b) of the five goals, whereas (2) and (3) are the indicators to measure progress in (d) and (e).

(1.1) Sustaining research activity levels

The first indicator examined was the number of publications per annum. Female researchers supported by the programme published 2.31 papers per year on average, which was compared with 0.63 papers from those who received no support. This is a 3.7 times greater output.

The second barometer was the number of patents won by female researchers (0.038 per person). The number was 0.025 per person for those who have not received any support.

The third indicator was the success rate of funding applications among female researchers with support (0.81 per head) and those without (0.25 per head). The likelihood of success was 3.2 times higher among those with support.

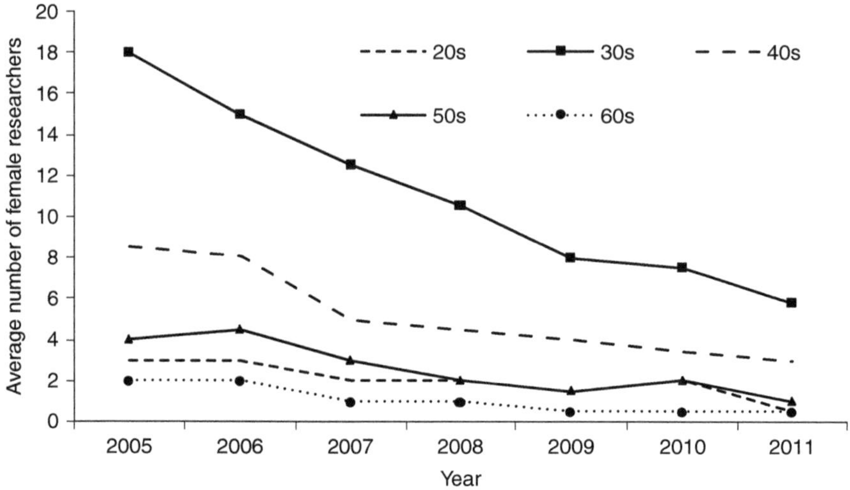

Figure 5.1 Average number of female researchers who resigned (per institution) among the MEXT-funded 'model' universities (by age).

Source: [15].

The average number of awards (per year, per person) among supported female researchers was 0.13, and in some engineering faculties where the ratio of women is very small, the number exceeded this score.

The evaluation report concluded that the MEXT programme did more than just sustain, but enhanced the research activities of female researchers during their maternity and child rearing period.

(1.2) Retention of female researchers

The average number of female researchers who left their workplace (i.e. 'model' universities) (per institution) has steadily declined since 2005.

Figure 5.1 evidently indicates sharp drops among female researchers in their 30s and 40s. The 'Model' programme proved successful in retaining female researchers in their maternity and child-rearing period.

(2) Increase of female staff members in the institution

Among the 55 'model' institutions, 2,333 female researchers were newly hired over a period of 6 years. This means an increase rate of 388.8 persons each year. The proportion of female researchers also increased at a higher rate than the national average, particularly from 2009 onwards, since the 'Acceleration' programme was introduced.

However, progress varies across STEM disciplines. The largest increase in proportion was found in health sciences (12 persons per institution) between 2005 and 2011. Yet the increase of 3.7 persons per institution in engineering is a remarkable achievement, given the low number of women in the field.

(3) Recruitment of female students into STEM fields

Out of 55 universities selected as 'model' institutions between 2006 and 2011, eighteen (32.7 per cent) saw an increase in the number of female students at any of the three levels (Table 5.5). Three universities, the increase took place across all three levels (undergraduate, postgraduate (Masters' and PhD levels)), while at another three universities, the increase was seen at postgraduate (Masters' and PhD) levels only.

As previously mentioned, 12 national universities were awarded the next-stage 'Acceleration' programme grant in 2009 and 2010, each for the duration of 4 years. Some progress was made at those universities, and they were noted in the evaluation report. However, at the time when this report was published, the programme was still ongoing.

The 'Model' project itself was not safe from criticism. Although the original aim was to increase women in STEM, it was altered to include all subjects from the year 2011 after some deliberation. Since the set-up of the programme in 2006, applications were made to the 'Special Coordination Funds for Promoting Science and Technology'. However, since 2011, this was re-classified as 'Science and Technology Human Resource Development Grants'. In addition, there was a criticism that while the Programme required a university-wide effort and investment, the funds only targeted STEM fields ([5]).

Additionally, the slow rate of increase in the number of female researchers in 'model' universities between 2006 and 2008 also attracted some criticism. Given the level of investment and frustratingly slow pace of results, the effectiveness of the policy was called into question. It was argued that the 'Model' programme was seen by some as funds to create a crèche or nursery on the university site, which became the be-all and end-all. These criticisms led to the creation of the 'Acceleration' programme in 2008. The 'Acceleration' programme also met with some opposition, both within selected universities and from other non-selected ones. After two rounds of selections, the programme had to be discontinued, despite the visible results of an increased number of female staff in those universities awarded. The politics of resource allocation is very much alive behind the scenes, while the strong focus on affirmative action to promote female staff has created some friction and encountered organisational resistance.

Table 5.5 Increase in the number of female students in the MEXT-funded 'model' universities

Increase in the number of female students at:	*Number of universities (N=55)*	*Percentage*
Undergraduate level	7	12.7
Postgraduate (MSc) level	15	27.3
Postgraduate (PhD) level	8	14.5

(*Source*: [15]. Adapted by the authors.)

However, support from MEXT, its agencies and the Cabinet Office has not wavered, and the basic policy goal and funding have been maintained. Since 2013, a new type of funding was created to fill the gap after the 'Acceleration' model was finished. The new 'hub and spoke' model for fostering institutional collaboration among academic and research institutes and industry (up to 1,500,000 JPY per year for the duration of three years). The 'Model' programme is still operational, and open to universities and research institutes that have not previously been funded.

From the SCJ surveys, there are clear indications that workplace cultures began to change in those universities and research institutes selected for the MEXT grant, which is a very positive sign. As social norms steeped in the national culture and workplace will require a long time to change, the results found in the selected institutions demonstrated the right direction.

5.2.3 EPMEWSE survey results

Chapter 4 described how various policies for supporting women in STEM were initiated largely as a result of the formation of the EPMEWSE, and its first survey. As with the SCJ, the EPMEWSE also carried out three large-scale surveys since 2003, and these data are invaluable sources, as the membership of each academic society affiliated to the EPMEWSE goes beyond universities and research institutes, encompassing postgraduate students, and researchers and managers in various firms. They reflect much wider views of women in STEM in Japanese society.

Table 5.6 shows the total number of respondents and the gender breakdown of each survey. Since the results of the first survey were reported in the previous chapter, this section focuses primarily on the latest survey results published in August 2013 ([6]), with some trend analysis using the second and third survey.

Female respondents accounted for the same 26.7 per cent in both second and third survey. This was an increase at more than 10 per cent from the first survey, and the actual number of female respondents also increased.

Table 5.6 EPMEWSE large-scale surveys

	Survey period	*No. of respondents*	*Male/Female*
1	August–November 2003	19,291	M: 16,147 (83.7%) F: 3,106 (16.1%)*
2	August–November 2007	14,110	M: 10,349 (73.3%) F: 3,761 (26.7%)
3	November–December 2012	16,314	M: 11,958 (73.3%) F: 4,356 (26.7%)

*38 respondents (0.2%) did not answer the question regarding their gender.

(*Sources*: [6], [16], [17]. Adapted by the authors.)

When we look at the whole range of respondents in 2007 and 2012, the ratio of respondents from the university sector increased by nearly 10 per cent (from 53.6 to 61.4), while the ratio decreased among respondents in the industry sector (from 23.2 to 14.8) in 2012. However, the female participants show a different trend, with the ratio of university researchers remaining stable and a greater number from enterprise taking part. The drops in the ratio were seen in research institutes ([6]: 13).

With regard to affiliations, in mathematics, physics and life sciences, 90 per cent of female respondents belong to a university or a research institute. In contrast, in electro-information, civil engineering and mechanical engineering, greater proportions of respondents belong to enterprise.

Concerning the age profile, the highest proportion of respondents are aged between 35 and 40 (28–30, 17.3 per cent), while amongst female respondents, the highest proportion was from a younger age cohort (24 years or younger, 36.4 per cent). Almost 60 per cent of female respondents have a PhD or equivalent, in contrast to 70 per cent of their male counterparts. Twenty per cent of both male and female respondents have a Master's degree.

The distribution of respondents by gender and their average annual income is shown in Figure 5.2.

In each age group, female respondents' wages were approximately 80 per cent of their male counterparts in all three types of affiliation (university, research institute and enterprise). Although this corresponds to the fact that women's positions were lower than men's in the same age category, the gender gap, rather than the affiliation gap, proved much greater.

The types of employment are in line with changes in Japanese society. Limited-term researchers for more than ten years have greatly increased between

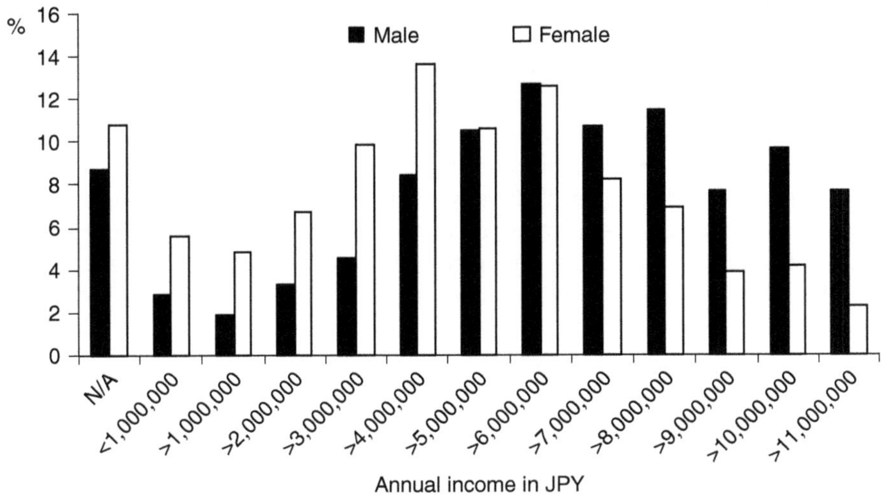

Figure 5.2 Proportion of respondents by annual income and gender.

Source: [6]: 15.

2007 and 2012. Almost 60 per cent of male and female researchers have experienced a job transfer or a change, or left their workplace. Reasons for those changes in workplace include 'career enhancement', 'end of contract', and 'contents of the job', while fewer female respondents named 'gender discrimination' than in 2007 ([6]: 101).

The question as to why they chose their current profession was asked. The results show similar patterns in 2007 and 2012. The top three answers were 'desire to search for truth', 'fulfillment of my capacities', and 'contributions to society', although a much higher proportion of male respondents chose these answers. Female respondents tend to believe there is less gender discrimination in STEM, and a good balance of work and life can be struck in their occupation.

Another question, 'Why are there few female leaders in STEM?' was also asked in the surveys. Again there was not much change between 2007 and 2012, in terms of the top categories. 'Difficulty of balancing work and private life' came first, followed by 'many job leavers and absence among women' and 'few women in leading positions at present'.

Men tend to agree with 'gender differences in aptitude for STEM', while women tend to believe that there is 'no desire for working under a female boss', and 'men are favoured in appraisal'. However, the latter two answers show a decline. Nearly 30 per cent of all the respondents (both male and female) in the 2012 survey carry the responsibility of caring for their elder parent/relative ([6]: 41). Social changes, including a rapid ageing population and more acceptance of women in higher positions, seem to be reflected in these results.

The awareness of various government policies among male and female respondents was examined. Figure 5.3 shows the proportion of female respondents who answered positively about the five major policies (the Restart Postdoctoral (RPD) fellowship (managed by the JSPS), 'Grant Scheme for Maternity and the Child Rearing Period' (managed by the JST), MEXT-led 'Model' and 'Acceleration' Programmes and 'Challenge Campaign' led by the Cabinet Office). Four of the five programmes (except the 'Acceleration' programme, established in 2009) started in either 2005 or 2006.

The JSPS's RPD Fellowship and the JST's Grant Scheme were far better known than the other three. Generally, there is greater awareness and higher approval rates among female respondents. The MEXT-initiated programmes were unknown to respondents, both men and women, while the Cabinet Office's 'Challenge' Campaign was supported by approximately one-quarter of both male and female respondents.

Lesser awareness of the MEXT-funded programmes was mainly due to the fact that they were targeted only at the university and research sector, and excluded industry. However, funding support for researchers during their maternity and child rearing period is clearly more popular, and perceived to be more effective than institutional support.

While analysing the effectiveness of JST's Grant Scheme, compared with RPD, Tatesawa (2014) identified four different types of support for female researchers in STEM by distinguishing purposes and target recipients (Table 5.7) [7].

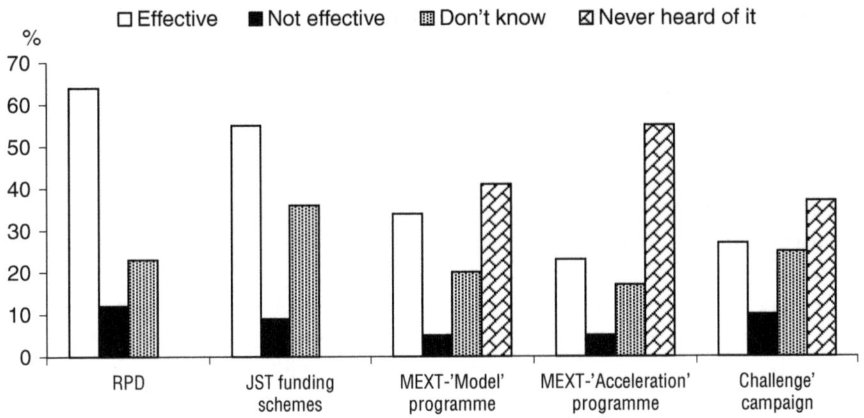

Figure 5.3 'What do you think about the effectiveness of these government policies?' (Female respondents).

Source: [6]: 49.

Table 5.7 Four types of support for women in STEM (by stage of career and purpose)

Primary purpose / *Target recipient*	*Continuation of research*	*Support for promotion*
Early-career researcher	Type 1	Type 2
Independent researcher	Type 3	Type 4

(*Source*: [7]: 4.)

> Type's 1 and 3: the primary purpose of this support is to help 'early-career' and 'independent' researchers continue their own research projects during their childbirth and child bearing period, while maintaining their employment status.
>
> Type's 2 and 4: this type of support is designed to move 'early-career' and 'independent' researchers to the next level in their careers. For example, for an 'early-career' researcher, it means a return to a full-time research position after maternity leave, and for an already 'independent' researcher, it is a career promotion during the childrearing period ([7]).

The JST Grant Scheme fits in with Type 1, as it is not directly paid to the researcher, but to the project team (Principal Investigator), and facilitates the change in culture within the research team and organisation. Tatesawa argues that although the PRD Fellowship (classified as Type 4) has improved the situation considerably, the JST Grant Scheme has a great role to play.

The findings from the EPMEWSE survey results show that setting the target proportion of female staff (quota system) proved divisive. Although more women are in favour of the system, 30 per cent of female respondents answered 'not effective' and more than 15 per cent identified some negative developments. Compared with the 2007 survey results, the ratio of people who responded 'the quota should be expanded' dropped sharply. Far fewer women (10 per cent, cf. 37 per cent in 2007) and men (4 per cent, cf. 15 per cent in 2007) support the expansion of a quota system ([6]: 50). This finding is corroborated by a different interview study with women in engineering ([8]).

Negative responses to setting the target ratio of female researchers came primarily from the younger male cohort (in their 30s), postdoctoral researchers and in the fields of physics and life sciences ([6]: 96).

The survey results show that the general patterns and challenges faced by female researchers remain the same, with some progress made in (perceived) gender discrimination. Although it is extremely difficult to replicate the level of impact that the first survey had on government policy, this large-scale survey still provides a snapshot of both male and female researchers in STEM fields at the time of the survey, and supplies invaluable longitudinal data.

This section has looked at three data sources (surveys in the university sector, the ministry's self-reflective review of its policy, and questionnaires conducted on members of academic societies) to gauge the developments and progress made in the situation of women in STEM in Japan. They highlighted the structural problems around wages and positions of female researchers, which have all persisted instead of disappearing. Yet gender aspects within the STEM workforce have clearly been considered more openly, and discussed and deliberated upon in the public domain. Although gaps still exist between public and private universities, among different disciplines and across different sectors, policies seem to have made an impact on frontline researchers.

The frustrating question of quantity (why so few?) has led to more drastic measures such as the 'Acceleration' programme and setting of the target ratio of female researchers by government. However, these policies divided opinions, and in some cases repeated the same mistake made in other countries such as Sweden (where women professors were considered by some to be second class, simply because they were selected using the quota system). Yet some other measures such as support for researchers have been positively perceived and proven effective.

The next section will present recent survey results, focusing more on the individual voices of female scientists and graduates who chose STEM subjects for their first degree in a women's university. It also highlights some emerging career progressions among women in STEM.

5.3 Voices of women in STEM

5.3.1 How is studying science perceived by women in STEM?

Major issues are facing women in STEM at the entry point of their higher education, and there is an absence of career options if they exit the labour market.

This section looks at survey results carried out at two women's universities to question why and when they chose STEM subjects as their specialisation, and how or whether they continued their careers.

5.3.2 Survey of alumnae of Tokyo Woman's Christian University

The university was founded as a women-only college with a Christian ethos. The very first admission exam in 1918 included three subjects (Japanese, English and Mathematics), which was unusual at that time, especially for a women's college. The motto of the college was: 'to foster an independently-minded human being with personality, equal to a man' ([9]). This was therefore slightly at odds with the notion of a 'good wives, wise mothers' which was commonly accepted in Japanese society at that time. Some authors argue that the overall aim of TWCU was not to foster professional career women, unlike other women's colleges such as Ochanomizu University (known as Tokyo Women's Higher Normal School) and Tsuda College ([10] [11]). However, in 1927, the government granted TWCU a special status, and the mathematics graduates were allowed to teach not only in girls' schools but also in boys' schools ([8]: 5).

After WWII, TWCU was upgraded to university status in the first wave of the education reform of 1948. However, its science department did not materialise until 1961, while mathematics was taught as part of liberal arts modules ([8]: 5). In 1968, the Department of Mathematical Sciences was officially established, and it now comprises natural sciences (physics, chemistry and biology) and information sciences (the latter was added in 2004).

A large-scale survey was conducted between January and April 2011. A questionnaire was sent to all the alumnae from TWCU's science section who graduated between 1931 and 2010, totaling 2,711 persons (out of 3,370 graduates on the register). The survey was conducted both in paper-based format and online. The response rate was 28.7 per cent (779 persons), of which 25 responses were not valid (N=754).

Given the history of the university, people aged 80 years old and over graduated from the former College (before TWCU became a 'university'). Also, at the time of this survey, those aged 68 years and younger are graduates of the Department of Mathematical Sciences.

Table 5.8 Valid responses by age group

Respondents' age group	*20–29*	*30–39*	*40–49*	*50–59*	*60–69*	*70–79*	*80 and older*	*Total*
Valid responses (N)	80	114	126	116	110	114	94	754
%	10.6	15.1	16.7	15.4	14.6	15.1	12.5	100.0

(*Source*: Kodate *et al*. 2014. Adapted by the authors.)

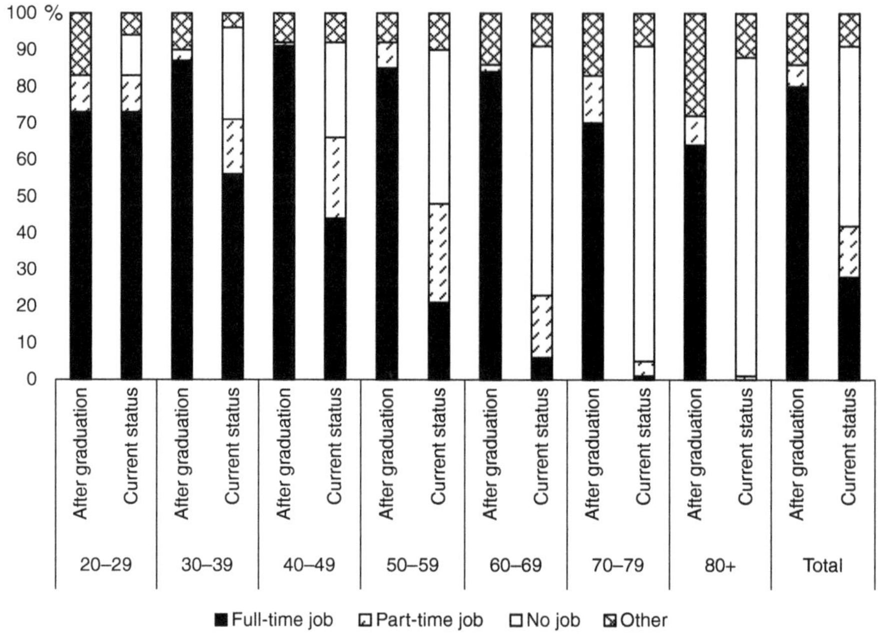

Figure 5.4 Occupation straight after graduation and current occupation.

Source: [8].

Figure 5.4 shows respondents classified by their current occupation types. Almost half (48 per cent) of the respondents are working (28 per cent full-time; 14 per cent part-time; 2 per cent self-employed and 4 per cent employer) (N=741).

As mentioned in previous chapters, the 1986 EEOL became a milestone in Japanese society. We also split the sample into two groups (those aged 50 years and over, and those younger than 49 years old), and conducted chi-square tests to examine if there is any significant difference.

Figure 5.4 shows the breakdown of employment status for each age group, both at the time of graduation and at present. The ratio of those who entered a full-time job was highest among those aged between 40 and 49. This coincides with the period after the EEOL was passed. But nearly half of them either quit the job or changed to part-time within approximately 20 years' time. However, nearly 70 per cent of the group are still working, which is not so different from the age group (30–39). Among those aged between 50 and 59, the ratio forming part of the working population goes down to 50 per cent.

When did you decide on your specialisation in science?

473 respondents (64.4 per cent) replied 'upper secondary school'. This is no surprise, given that students often make up their minds on specialisation based on their strengths in subjects in the run-up to rather competitive university entrance exams.

Table 5.9 When did you decide to specialise in STEM?

Respondents' age group	*20–29*	*30–39*	*40–49*	*50–59*	*60–69*	*70–79*	*80 and older*	*Total*
Kindergarten & Primary school	18.8%	11.5%	19.2%	10.5%	3.8%	6.4%	4.5%	10.7%
Lower secondary school (Junior High)	32.5%	31.0%	23.2%	28.1%	13.2%	14.7%	21.6%	23.3%
Upper secondary school (High)	48.8%	56.6%	57.6%	61.4%	83.0%	76.1%	64.8%	64.4%
After Secondary school	0%	0.9%	0%	0%	0%	2.8%	4.5%	1.1%
Other	0%	0%	0%	0%	0%	0%	4.5%	0.5%
Total number	**80**	**113**	**125**	**114**	**106**	**109**	**88**	**735**

(*Source*: [8]. Adapted by the authors.)

However, when the two groups (50+ and 20–49) were compared, the younger cohort tends to decide earlier, before entering the upper secondary school ($\chi^2 = 31.9$; $P < 0.05$, odds ratio (OR): 2.4, 95 per cent confidence interval (CI): 1.773.26).

What influenced your decision to enter STEM fields?

School teachers at primary, lower and upper secondary levels seem to have the greatest impact, irrespective of age group. Among the respondents in their 40s and 60s, science classes and experiments were mentioned more frequently than the rest, while people in their 20s chose 'family and relatives' and 'books and media'. 'Other' included 'cram school teacher', 'I liked science', and 'I was very good at science at school'.

Family influence

The survey asked the question 'in your family, is there anyone who specialised in STEM?' Apart from those aged 80 and older, approximately 40 per cent of the alumnae had at least one parent who had studied STEM.

When looking at the ratios of those whose mother also studied STEM subjects, a significant difference was found between the two groups (50+ and 20–49). The younger cohort had much higher proportions of mothers in STEM (8.7 per cent for the 40s, 7.9 per cent for the 30s and 15 per cent for the 20s) ($\chi^2 = 12.3$; $P < 0.05$, OR: 2.9, 95 per cent CI: 1.56–5.39). When grandparents' STEM specialisation is recognised (at least one grandparent studied STEM), the average ratio of people is 10.7 per cent. A higher proportion was found among those in their 30s and 20s (17.5 per cent and 16.3 per cent, respectively).

Are you happy with your choice of STEM?

Out of 719 respondents, 684 (95.1 per cent) replied positively to this question above. People who are extremely satisfied account for 55.4 per cent (398 persons), and these figures are higher for age groups younger than 50.

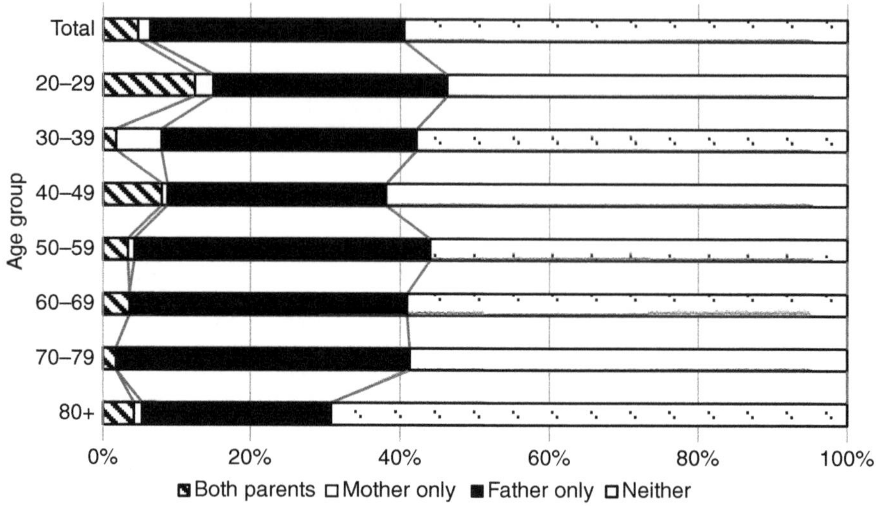

Figure 5.5 'Did your parents also specialise in STEM?'

Source: [8].

Top three reasons were: (1) It was my favourite subject, (2) studying science strengthened logical thinking, and (3) I had the advantage in applying for a job. Other reasons included 'I obtained a licence as a teacher', 'I felt I had an added value as a woman in science (rare in society)', and 'it was useful in everyday life'.

Some of the negative reasons included the following. 'It can be no use unless you become a specialist in the field'. 'I was good at sciences as an exam subject at high school, but once I entered university, the same style of learning did not work' ([8]: 24).

Would you recommend choosing STEM to the next generation, including your own children?

Out of 605 valid responses, the positive answer accounted for 78.0 per cent. There was not much difference by age group, apart from those aged 80 years and over.

Table 5.10 Would you recommend choosing STEM to your children or the next generation?

	Yes	*No*	*No Response*	*Total Number (N=605)*
20–29	70.0%	21.3%	8.8%	73
30–39	75.4%	11.4%	13.2%	99
40–49	65.9%	21.4%	12.7%	110
50–59	62.9%	15.5%	21.6%	93
60–69	56.4%	18.2%	25.5%	82
70–79	64.0%	15.8%	20.2%	91
80 and over	39.4%	21.3%	39.4%	57
Total	**62.3%**	**17.6%**	**20.0%**	**100%**

(*Source*: [8]. Adapted by the authors.)

The most popular positive answer was 'because studying science fosters logical thinking'. Among the younger cohort, 'there are opportunities, particularly for women in STEM, because there are not many' was also popular ([8]: 29).

Overall, the survey reported a very positive experience of graduates specialising in mathematics and other science subjects at this university. Their decision to choose STEM seemed to be influenced primarily by their family and school environment, and their own maternal influence among the younger generation merits further attention and research.

5.3.3 Survey of alumnae of Japan Women's University

The Faculty of Science at JWU conducted a survey in 2007 as part of their MEXT-funded 'Model' programme. One thousand eight hundred and thirty-five alumnae who had graduated from the Faculty of Science (formerly Faculty of Home Economics, science divisions) since the early 1980s were sampled, and sent a questionnaire. Three hundred and sixty-seven (20 per cent) responded.

Of those 367 respondents, 317 (86.7 per cent) have a Bachelor degree in Home Economics (until 1992) or in Science (1993 onwards). Twenty-six (7 per cent) have a PhD, while 43 (11.7 per cent) obtained MSc or equivalent.

This survey shows a result which is very similar to that of TWCU in that 288 respondents (78.4 per cent) are satisfied with the fact that they specialised in STEM. Seventy-six per cent have a partner, and 64.5 per cent have at least one child. What was striking was that nearly 60 per cent (217 respondents) balance work and private life with children (45.8 per cent in full-time employment, 13.4 per cent in part-time). Of those who work, 38.7 per cent work in private enterprise, 3.8 per cent are employed in public research institutes, and 12.8 per cent work in the university and education sector. More than half (53.2 per cent) responded that they use knowledge and skills learned in higher education, and would continue their work.

In 2011, a follow-up survey was conducted, only targeting 90 graduates from a particular physics group. The response rate was 50 per cent (N=45).

Thirty-six out of 45 are in full-time employment (80.4 per cent), and the largest group (11, or 24.4 per cent) works in a technical development unit in a private company, followed by a research unit, also in a private company (9, or 20 per cent). Six respondents work in the service sector, while five respondents are university lecturers. Although 12 out of 45 (26.7 per cent) have changed jobs at

Table 5.11 Valid responses by age group (JWU survey (1))

Respondents' age group	*28–29*	*30–34*	*35–39*	*40–44*	*45–47*	*Total*
Valid responses (N)	14	73	95	139	46	367
%	3.8	19.9	25.9	37.9	12.5	100.0

(*Source*: [18]. Adapted by the authors.)

Table 5.12 Valid responses by age group (JWU survey (2))

Respondents' age group	*21–25*	*26–29*	*30–34*	*35–39*	*40–44*	*45–49*	*50–54*	*55–59*	*Total*
Valid responses (N)	6	13	4	8	7	1	3	3	45
%	13.3	28.9	8.9	17.8	15.6	2.2	13.3	13.3	100

least once, a very high proportion of the graduates is continuing to work. Four out of five respondents among those aged 30 years and over are married, while half of them have a child or children.

They listed what they regarded as the key factors to being able to continue working. The top three were: partner and family's understanding and cooperation, boss and male colleagues' support at the workplace, and individual effort and pride. Positive action, change in women's way of thinking, and change in the socially accepted value system were also mentioned. With regard to primary reasons for the low number of women in STEM, a majority of respondents referred to educational environment, the socially accepted value system ('men at work, women at home'), and few role models.

These findings are generally in line with other survey results by EPMEWSE and other universities (e.g. Yamagata University). One area on which the split between 'yes' and 'no' respondents diverges so greatly in the surveys can be summed up by the statement 'husbands should work outside, while wives should keep the home'. In the 2012 survey conducted by the Cabinet Office, 51.6 per cent of the respondents agreed with the statement ([12]). The proportion of those who said yes increased recently for the first time, even among the young generation, after having decreased steadily between 1992 and 2009. However, the results from another survey conducted in a national university in 2011 show only 22.2 per cent of male respondents and 13.5 per cent of female respondents agreed with the statement (N=1500) ([13]).

Just as the opinion polls, the images portrayed by these survey results can vary greatly, depending on the target participants. This example shows a potentially widening gap between people located in a university setting and the public in general. In fact, at individual level, there are now a number of active female researchers in STEM, which the next section sheds light on.

5.4 Emergence of new STEM women and their career paths

5.4.1 Eight multi-career paths

This section illustrates the career paths of eight female graduates. They all have a bachelor degree or higher, and are currently in full-time employment. They are still using the skills and experience acquired through their period in higher education.

[*Undergraduate degree*] After completing their bachelor degrees (all in science), the first four persons ventured into different areas.

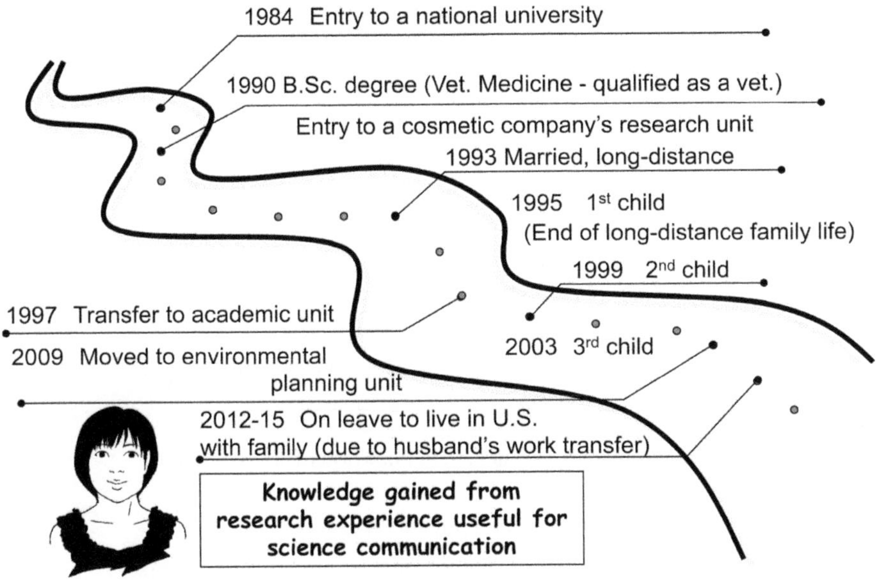

Figure 5.6(a) Ms A. Changing track from research to science communication in a private company.

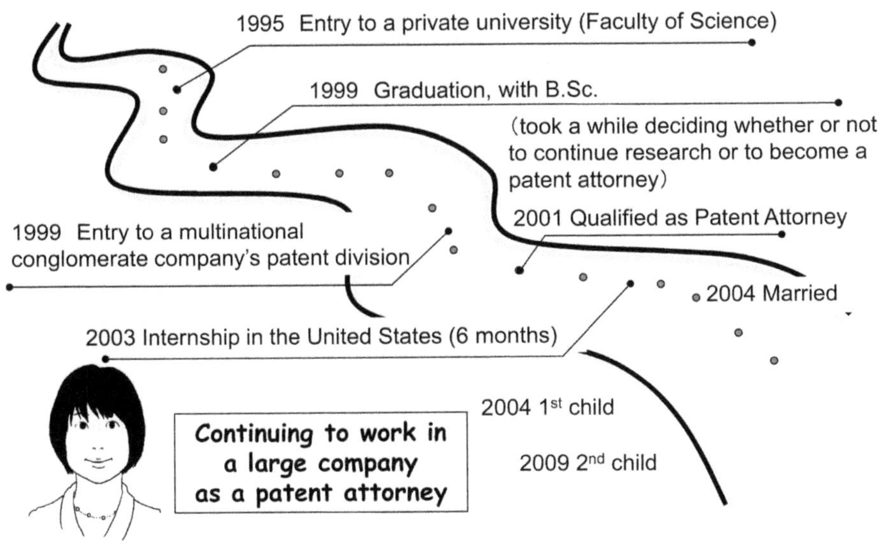

Figure 5.6(b) Ms S. From BSc to patent attorney.

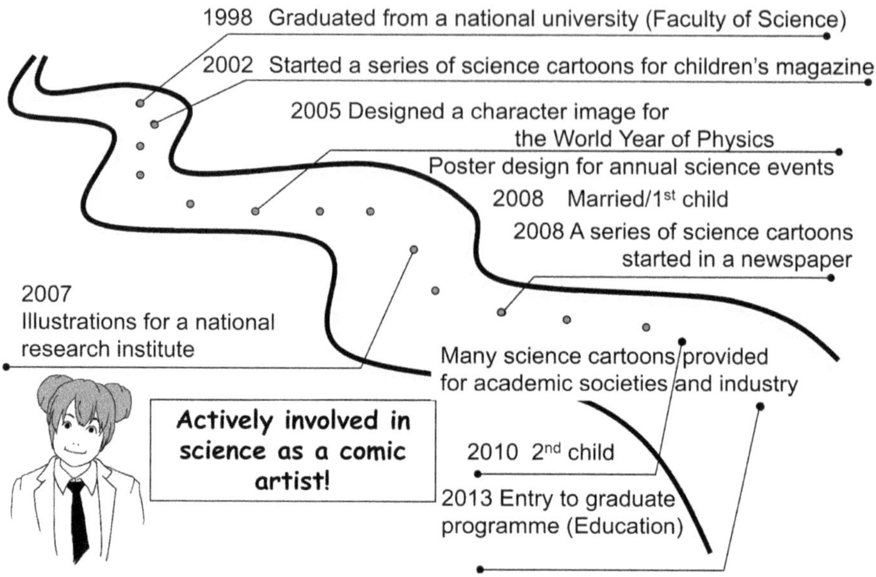

Figure 5.6(c) Ms W. From a physics degree to the world of science comics.

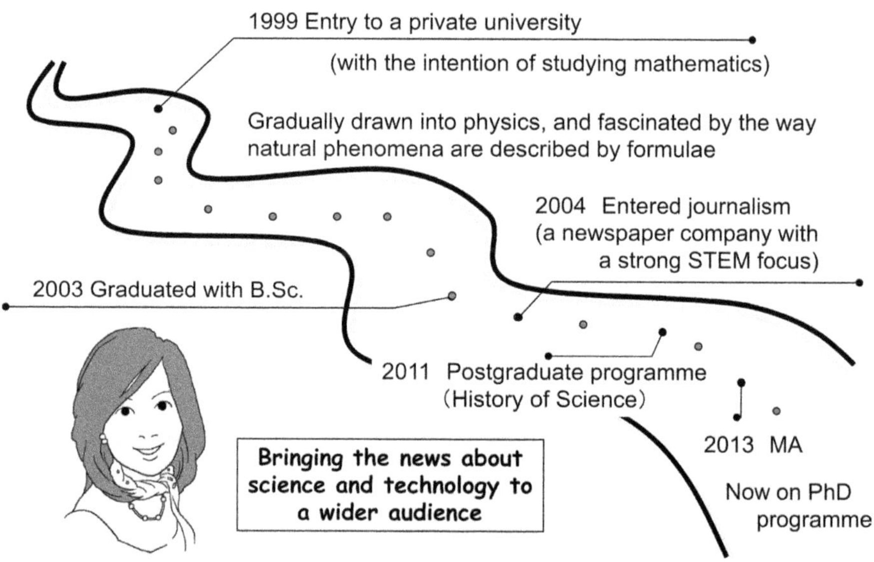

Figure 5.6(d) Ms F. From BSc to postgraduate studies, via a career in journalism.

Source: Designed by the authors and Hayanon.

[*Postgraduate degree*] Dr O. followed the most traditional model, graduating from a woman's university and entering a national university to pursue postgraduate degrees. Dr Y. even went on to change her career path quite radically, and became a high school teacher. This would have been the normal path to follow 100 years ago for women in STEM, but these days, it is not. In contrast, Dr F. went through four short-term contracts before becoming a full-time university lecturer.

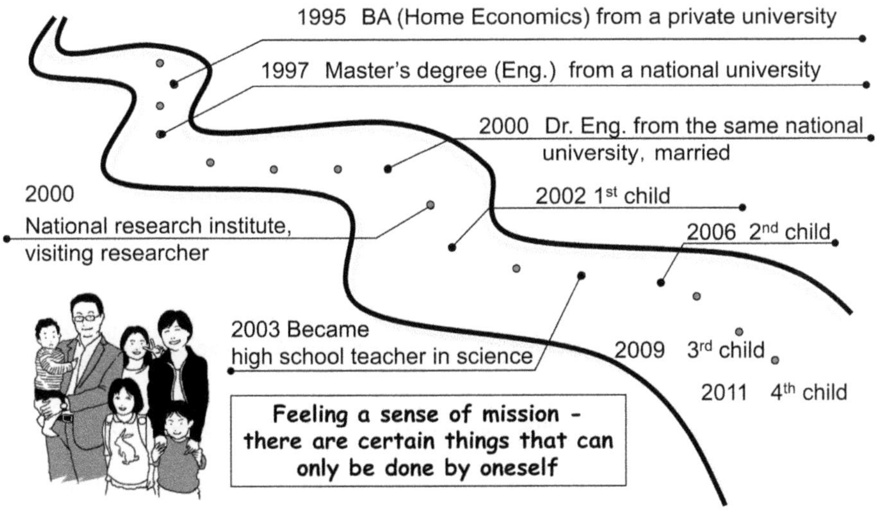

Figure 5.7(a) Dr O. From research to high school science teacher.

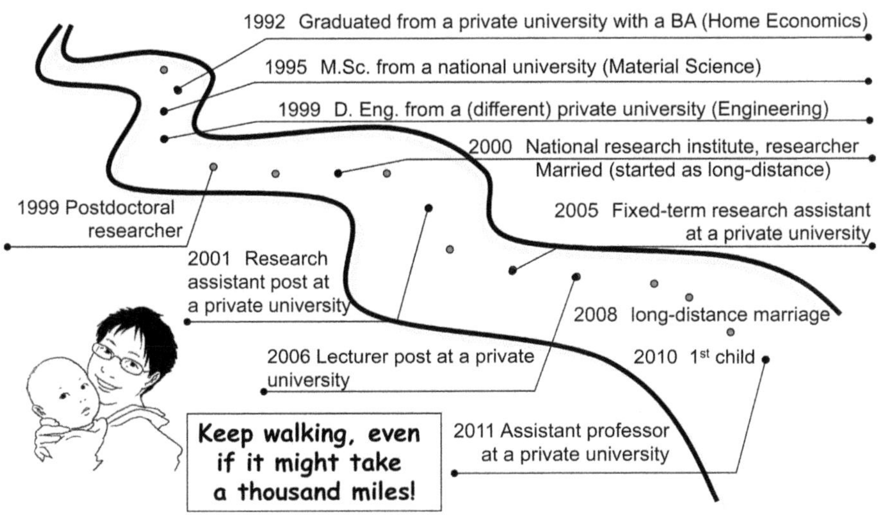

Figure 5.7(b) Dr F. From fixed-term contracts to a full-time university teaching post.

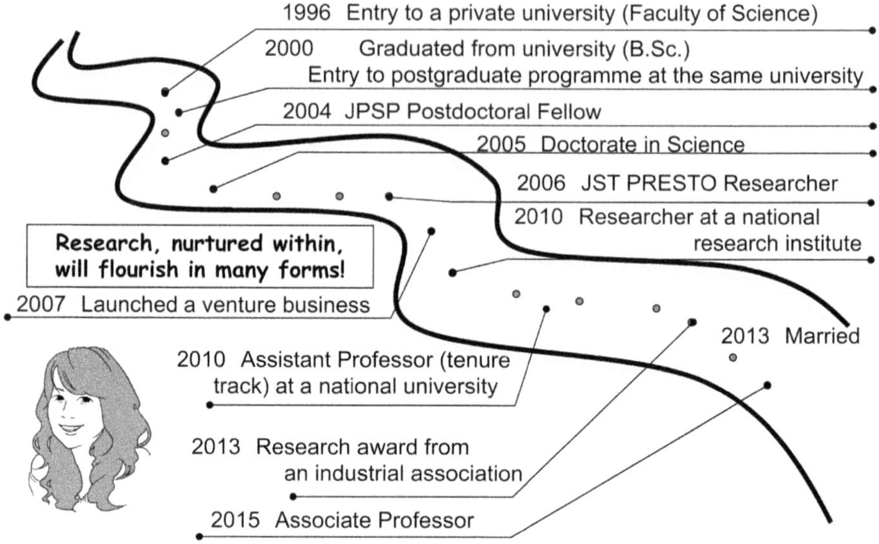

Figure 5.7(c) Dr W. From research and venture business to Associate Professor

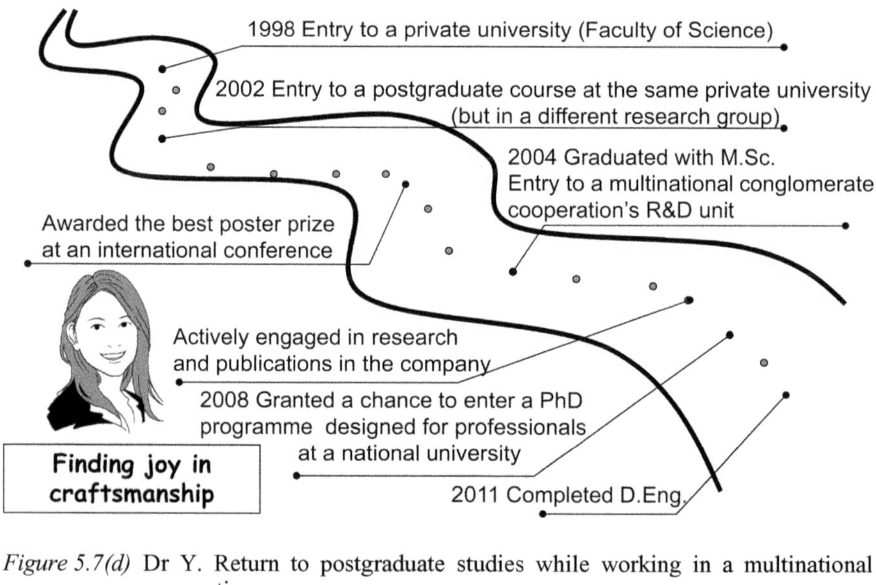

Figure 5.7(d) Dr Y. Return to postgraduate studies while working in a multinational corporation.

Source: Designed by the authors and Hayanon.

Dr W. is also a new type of female researcher in STEM: completing all her studies at the same private university, successfully obtaining a research fund, creating a business venture and then becoming a lecturer at a national university. On the other hand, having finished her Master's degree, Dr Y. pursued her research career in a large company's R&D unit, and managed to obtain support from the company to undertake and complete her PhD.

Although this is not a comprehensive list of STEM women's new career paths, these illustrations shed light on the manifold possibilities open to them.

5.4.2 Role models who have research experience overseas

There are also women in STEM who have extensive research experience overseas, and have therefore broken the mould. In this category, two types of STEM women exist: (1) obtaining a PhD overseas, and (2) working as a researcher abroad.

Out of 31 researchers featured in JST's publication, *Role Models: a bright future for women in science, engineering and technology* ([14]), 12 researchers in STEM have either obtained a PhD abroad or worked outside Japan for more than one year (1 in mathematics, 4 in physics, 3 in chemistry, 1 in life sciences and biology, and 3 in engineering). Their biographies and observations about the difference between Japan and the country in which they researched are summarised in the Table 5.13.

5.5 Conclusion

This chapter has presented data to highlight some of the changes that have happened since a variety of policies were introduced in the mid-2000s. The large-scale surveys conducted by SCJ and EPMEWSE are helpful in tracking and monitoring these changes. They show some signs of change in the workplace culture and in society.

The evaluation report of the MEXT-funded 'Model' and 'Acceleration' programme emphasises the important changes and the shift in the right direction in terms of the increase in the number of female staff. However, these programmes are not widely recognised by the members of academic societies, and affirmative action appears to be seen as ineffective by those members.

There are other policies targeting individual researchers and research teams, which are also being reviewed, and they have resulted in some positive results. Compared with the 'Model' programme, the results may be on a much smaller scale. However, when persistent cultural and societal norms and values must be dealt with, the heuristic approach (applying manifold policies) should be the only way forward.

On an individual level, the majority of those who specialised in STEM subjects are satisfied, and firmly believe that their learning was valuable to themselves and society. A variety of career paths are now open to many female graduates with STEM skills, and in fact, there are many women who are active in various fields, using their experience and knowledge, while balancing their private life and work. This in itself is something to be celebrated in society, given the length of time it has taken before this became possible and even seen as normal in Japan. Although there are still a number of obstacles and challenges, there are many opportunities as well, and the time has come to fulfil one's potential. Women in STEM are no exception to this.

Table 5.13 Women in STEM who have a career and research experience outside of Japan

	Area of research	*Age group*	*Career path*	*Crucial factor for career decision*	*Insights obtained from experience overseas, and noted differences from Japan*
1	Condensed material physics	40s	National university (women-only) → National University (PhD) → Research Fellow in the USA → Professor at National University	Prior to my university entrance exam, there was a period when I was unable to decide between many subjects, including physics, mathematics, English, and medicine. In the end, my curiosity about natural phenomena and a desire to have an experience, not possible during study at my desk, led to experimental physics.	There were many female researchers in the research institute, and they were full of life at work. I did not find much difference in research life between men and women overseas.
2	Particle physics	40s	National university (women-only) → Researcher (tenure) in France	No doubts about my choice of STEM from the onset. Interested in the workings of a mystery that looks natural at first sight.	In this field, there are only a few women in any country. More women work after childbirth in France than Japan.
3	Astronomics	30s	National university (women-only) → PhD in the USA → Researcher in astronomical observatory in the USA.	At the time of the university entrance exam, I chose physics based on my own interest. My parents and teachers were worried about my choice to specialise in science. When deciding to study abroad, I changed my specialty to astronomics.	In academia, there are few women in any discipline. Yet half of the researchers in my current observatory are female, and the deputy head is also female. While there is no discrimination, there is no special treatment either!
4	Engineering (Atom Technology)	50s	Private university (PhD) → researcher at national institute in Japan → with partner, went to Canada, and became postdoc researcher → back in Japan, senior manager in a large company.	I was fascinated by the mystery of natural phenomena during physics experiments at high school.	As a woman, freedom of thought (in Canada), which does not exist in Japan, was a great difference. Liberated from the shackles of a sense of self, for the first time I could see many important things and their essence.

5	Mathematics	50s	National university (women-only) (MA), qualified as teacher. When married, went to France, and became PhD student on government scholarship → back in Japan, professor at private university.	Maths teacher in my high school	In France, there are many female mathematicians (perhaps 25 per cent). It is difficult to gain insight into international academic standards if one stays in Japan.
6	Material science	60s	National university (women-only) → national university → Postdoc in the USA → professor at national university.	Preparing for the university entrance exam, I thought of literature, but my parents' objected, telling me to acquire expertise. I chose chemistry, as I was fascinated by experiments.	Being in the USA in the 1980s, I experienced first-hand affirmative action, and formed a network with female researchers. Later on, they became leaders in the field, which inspired me.
7	Material science	50s	National university (women-only) (MA) → research assistant until getting married → with partner, went to the USA and became researcher → back to Japan, obtained PhD from national university → professor at private university.	Grew up in a house where my grandfather and father also specialised in chemistry. Through the route of home economics to chemistry. Balancing research and family life, bringing up children in the USA for 15 years led to a life as a researcher.	Upon entering the graduate school in Japan, I was asked if I had a fiancé. That is unthinkable in the USA. Instead of thinking of myself as a 'woman, wife and mother', I had a serious thought about what I would like to be as an 'individual human being'.
8	Oceanography	40s	National university (PhD) → university in the USA. Moved to Canada and became researcher → back in Japan, now associate professor at national university.	Scenery I saw as a primary school child in Norway brought me back to a northern country again!	Great surprise to find so many female researchers on board in Canada. National university in Canada showed high sensitivity towards diversity, and balance between work and private life. Project leader was also female.
9	Molecular embryology	60s	National university (MSc) → researcher in a company's research institute → Married, went to the USA and became researcher. → Back in Japan, obtained PhD (agriculture) → professor at private university.	No doubts about my choice in science from the outset. I selected the university where I can specialise in biochemistry.	I was inspired by taking part in a symposium on 'transposon' in 1980, where all the leading figures gathered from across the world. I experienced affirmative action in the USA at first hand. Two female associate professors were so powerful that their voices had impact on the faculty.

(Continued)

Table 5.13 Women in STEM who have a career and research experience outside of Japan (Continued)

	Area of research	*Age group*	*Career path*	*Crucial factor for career decision*	*Insights obtained from experience overseas, and noted differences from Japan*
10	Civil engineering	50s	National university (civil engineering) (BA) → entered architectural firm, married and quit → with partner, went to Thailand, as a housewife → later obtained PhD in the UK → Back in Japan, associate professor in national university's research institute.	In spite of advanced science and technology in Japan, natural disasters overwhelm the country. I wanted to contribute to changing this.	My study aboard changed my career as a civil engineer to a researcher. Found liberation in the way people's performance was evaluated irrespective of age and educational background. Slower passage of time than in Japan allowed me to ponder things. Both men and women had a better balance of work and private life.
11	Nuclear engineering	50s	National university (BEng and MEng) → PhD in the USA → Back in Japan, professor at national university.	As a primary school child, I was struck by the power of science (Apollo11), and the necessity to understand it (oil crises).	Even in the late 1980s, 30 per cent of students were female! There were differences between women in an academic society in the U.S. and those in Japan. Although I did not believe it is possible to have both work and family life, I changed my mind by seeing how the Americans do it.
12	Mechanical engineering	40s	National university (MEng) → entered a large company → Obtained PhD from private university → Postdoc in Italy → Back in Japan, assistant professor at national university.	I originally thought about entering medicine, but watching an artificial arm on TV interested me in medical engineering. Before the university entrance exam, I thought about studying humanities subjects, too.	I discovered a good balance between work and private life, and learned how to see Japan from a wider perspective. There are not so many women in mechanical engineering in Italy either, but Japan has even fewer. For good or bad, women in Italy are not as conscious of their gender as those in Japan.

(*Source*: [14].)

The next chapter will conclude the book by looking at present and future challenges and opportunities for Japanese women in science and engineering.

References

[1] Science Council of Japan. Gakujyutsu bunya ni okeru danjyo kyōdō sankaku sokushin no tameni (For the Promotion of Equal Participation of Men and Women in Academia). [In Japanese]. [online]. Available from: http://www.scj.go.jp/ja/info/kohyo/pdf/kohyo-20-t60-8.pdf, 2008. Accessed 26 October 2014.

[2] Science Council of Japan. Nihon no tenbō – gakujyutsu kara no teigen 2010 (Prospects for Japan: Recommendations from Academia) [In Japanese]. [online]. Available from: http://www.scj.go.jp/ja/info/kohyo/pdf/kohyo-21-tsoukai.pdf, 2010. Accessed 26 October 2014.

[3] Science Council of Japan. Gakujyutsu ni okeru danjo kyōdō sankaku suishin no kasoku ni mukete (Towards Acceleration of Promoting Gender Equality in Academia). [In Japanese]. [online]. Available from: http://www.scj.go.jp/ja/info/kohyo/pdf/kohyo-21-h133-2.pdf, 2011. Accessed 27 October 2014.

[4] Ministry of Education, Culture, Sports, Science and Technology. Kagaku gijyutsu kihon keikaku: Kakugi kettei (The 4th Science and Technology Basic Plan). [In Japanese]. [online]. Available from: http://www.mext.go.jp/component/a_menu/science/detail/__icsFiles/afieldfile/2011/08/19/1293746_02.pdf, 2011. Accessed 17 November 2014.

[5] Former Director, Gender Equality Office, Japan Agency of Science and Technology. Personal communication.

[6] EPMEWSE. Dai sankai kagaku gijyutsu kei senmon shoku no danjo kyōdō sankaku jittai chōsa (The Third Survey on Gender Equality among Professionals in Science and Technology) [In Japanese]. [online]. Available from: http://www.djrenrakukai.org/doc_pdf/2013/3rd_enq/3rd_enq_report130918.pdf, 2013. Accessed 17 November 2014.

[7] H. Tatesawa. JST shussan/kosodate shien seido ni kansuru kōsatsu (Discussion on Support for Funding Childbirth and Childcare of the Japan Science and Technology Agency) [In Japanese]. Unpublished Technical Report, 2014.

[8] T. Kodate, N. Oguchi, N. Kodate, and N. Takahashi. Joshi daigaku rikei gakka no jyūyōsei nitsuite. Technical report, Tokyo: Institute for Women's Studies, TWCU, 2014. (Final Report: Women's Universities and the Importance of their Role in Encouraging Women in Science).

[9] Tokyo Woman's Christian University's 50-year History, Book Editing Committee. *Tokyo Joshi Daigaku gojyūnenshi (Tokyo Woman's Christian University's 50-year History)*. [In Japanese] Tokyo: TWCU Institute for Women's Studies, 1968.

[10] M. Amano. *Jyoshi kōtō kyōiku no zahyō (Perspectives on Girls' Higher Education)*. [In Japanese] Tokyo: Kakiuchi Shuppan Publishing Ltd, 1986.

[11] N. Mamatkulova. Senzenki jyoshi kōtō kyōiku no kyōiku rinen ni kansuru ichi kōsatsu: sono tokushitsu to Tsuda Eigo Jyuku no ichiduke wo chūshin ni (the Ideals of Women's higher education in prewar times: Focusing on their characteristics and placing "Tsudaeigakujuku" in the history of women's higher education) [In Japanese]. Technical Report 2, *The Bulletin of the Graduate School of Education of Waseda University*, 2012. [online]. Available from: https://dspace.wul.waseda.ac.jp/dspace/bitstream/2065/35637/1/KyoikugakuKenkyukaKiyoBetsu_19_2_Mamatkulova.pdf. Accessed 31 October 2014.

[12] Cabinet Office. Chōsa kekka no gaiyō, (Overview of the Survey Results) [In Japanese]. [online]. Available from: http://survey.gov-online.go.jp/h24/h24-danjo/2-2.html, 2012. Accessed 17 November 2014.

[13] Office for Gender Equality, Yamagata University. Yamagata work–life balance innovation dai nibu, Heisei nijyūni nendo danjo kyōdō sankaku ni kakawaru ankto kekka hōkokusho (Survey Results Regarding Gender Equality–Yamagata Work–Life Balance Innovation Part II (2010)). [In Japanese]. [online]. Available from: http://www.yamagata-u.ac.jp/kenkyu/danjo/newsletter/pdf/pdf20100331/dai2buanketo.pdf/y21report2(full).pdf, 2011. Accessed 17 November 2014.

[14] Japan Science and Technology Agency. *Rikei jyosei no kirameku mirai* (Role Models: A Bright Future for Women in Science, Engineering and Technology) [In Japanese]. (3rd Edition). Tokyo: JST, 2011.

[15] Ministry of Education, Culture, Sports, Science and Technology (2012) *Kagaku gijutsu shinkō chōseihi*: *Program hōkokusho* (Evaluation Report on Special Coordination Funds for the Promotion of Science and Technology). [In Japanese] [online]. Available from: http://www.mext.go.jp/a_menu/jinzai/hyouka/__icsFiles/afieldfile/2013/01/25/1329874_03_1.pdf. Accessed 28 October 2014.

[16] EPMEWSE. *Diverse Visions of Scientists and Engineers in the 21st Century: For the Promotion of Gender Equality*. Technical report, Report on the MEXT-commissioned project, 2003, 2004.

[17] EPMEWSE. *Kagaku gijyutsu kei senmon shoku niokeru danjo kyōdō sankaku jittai no daikibo chōsa* (The Survey on Gender Equality Among Professionals in Science and Technology) [In Japanese]. [online]. Available from: http://www.djrenrakukai.org/2007enquete/h19enquete_report_v2.pdf, 2008. Accessed 17 November 2014.

[18] Japan Women's University. *Saiensu ni idomu jyosei tachi* (Women in Science Who Take on the Challenge). [In Japanese]. Tokyo: Ad Three Publishing, 2009.

Conclusion

C.1 Past and present: how far have we come?

This book has traced the history of Japanese women in science and engineering, with a particular focus on recent policies and initiatives to support them. It has emphasised the significance of the changes that took place over time. Some were sudden and formal policy changes, due to dramatic historic events such as the end of WWII, or external pressure, while others were more gradual and informal. Like many cultural shifts, it required a long period of time to see the impact of these policy reforms on society and individual behaviours. Institutionally, Japan kept the post-WWII regime under strong government, underpinned by a centralised civil service and one-party rule, up to the early 1990s. The regime became a highly effective vehicle for economic growth and the standardisation of national life. Until 2010, Japan's economy had been the second largest in the world for 42 years, and the nation once achieved the status of an 'all middle-class' society. However, it has had its own weaknesses, particularly in adaptability to social change and the capability to embrace diversity and uncertainty.

Policy changes that occurred in education and the labour market were often met with fierce (albeit invisible at times) resistance on the ground. They took various forms of institutional adaptation, be it the creation of two-year colleagues for women or the emergence of a two-track system in workplace after the EEOL, to maintain the traditional path. Therefore, in the areas of STEM, it has been entirely up to individual courage and informal efforts, first to choose a STEM programme and then to remain committed to working in research, education or industry up until very recently. This is still true in many cases, but a woman's decision to specialise in science and engineering, and continue pursuing a career in those fields is now officially supported in Japan. This is a dramatic change.

With hindsight, the policy change in the mid-2000s appears to be natural progression from previous events such as the enactment of BAGES, and the Science and Technology Basic Act. However, changes in the previous government structure, the formation of EPMEWSE (with membership from both academia and industry), and the realisation that Japan was far behind other Western countries in the advancement of women in society were the three key factors that enabled the reform to take place. A broadly shared sense of crisis that existed (exists) has also pushed the agenda of gender equality forward. The country needs to tackle low fertility and a declining population, while sustaining its economic growth and world reputation as a high-tech nation. What was different from policy changes in the past was the role women played themselves at different stages in the policy process. Even though the proportion of female leaders in society as a whole is low, in the early 2000s, women were holding some key positions. The then Minister of Education and senior civil servants (mostly female) were providing full support for the development and implementation of the policies. Under the banner of 'equal participation of men and women', a number of male leaders were also involved in the processes and publicly endorsed the changes.

Nearly ten years after the policy change, these programmes and initiatives have been evaluated, and tweaked several times, having met some criticisms and backlash. However, the government's commitment to utilising women's power and knowledge pool for academic and scientific innovation has not swayed, but gained momentum, at least in the rhetoric. The commitment has been so far matched by their continuation of various programmes. Therefore, it is safe to conclude that the level of formal support has been established as some form of consensus within the policy and academic community. Nevertheless, given the long history and the persistence of social norms, this may only signify that only highly qualified and elite women get close-to-equality treatment, leaving others behind [1].

As seen in the Introduction, the latest government proposal to parliament has been dubbed the 'Advancement of Women' bill. However, its over-emphasis on economic growth, targeting women as a precious source of labour and skills has been critically viewed ([2] [3]). Without proper investment in implementing women- and family-friendly programmes, there is the danger that Japan could swing from a 'male-breadwinner' model to a 'dual-earner/female double-burden' model, rather than the simple 'dual-earner' model, found in Scandinavia and other Northern European countries ([4] [5]). Hiroki Yoshida, Managing Director of a non-profit organisation Fathering Japan (founded in 2006) argues that what is needed is a cultural shift in men's mentality rather than women's. Others also criticise the fact that the focus has now been shifted from 'equal participation of men and women' to just 'women'.

The gender gap in wages is still an issue highlighted by the NTA survey ([6]). When including those with only temporary or part-time employment, income among female workers was only 53.3 per cent of males'. The figure has remained constant at around 50 per cent since the 1980s. The majority of women leave work due to pregnancy and childbirth, but even for those who return to work, employment status tends to be on a part-time or temporary basis ([7]).

Furthermore, in recent years, economic inequality has increased in Japan, which is causing social problems such as poverty among young parents and children including single mothers ([8]). In fact, Japan has the fourth highest level of relative poverty (based on the number of people who live on less than half the median income) after Mexico, Turkey and the USA. ([9]). Additionally, a recent report indicates widening inequalities between urban and rural areas in the enrolment rates of high school students at higher education institutions. The gap has doubled in the last 20 years ([10]).

These are some of the present challenges facing the Japanese in general, including women in STEM. Yet the greatest achievement so far is that many women are now able (and have indeed managed) to diversify their career paths and utilise their own potential to the full, as demonstrated in Chapter 5, if they want to. Many of the previous formal barriers have now been eliminated. Yet further efforts should be made to ensure equal opportunities will be provided for children from an early age, students from different socio-economic status and geographical locations.

The remainder of this concluding chapter looks at recent developments surrounding Japanese girls and women in STEM, and future challenges and opportunities.

C.2 Recent developments, future challenges and opportunities

Despite recent scandal surrounding a researcher in RIKEN (see Introduction), the damage of that event on girls' choice of STEM subjects seems minimal. The increased popularity among girls for entering science and engineering faculties was recorded at the 2014 entrance exam to universities. This meant that competition to get into those faculties became more intense for both boys and girls ([11]).

This report was endorsed by personal communications which took place between senior staff members of three girls' high schools (e.g. principals and science teachers) and the Gender Equality Bureau of JST ([12]). These three high schools, all based in Tokyo, are well-known for their high academic achievements and having nurtured female leaders in many fields. In all three schools, a greater number of students chose STEM over humanities subjects in recent years. The total ratio of students electing natural science subjects (as opposed to human sciences) at the university entrance exam ranges from 40 to 70 per cent in those schools. This is much higher than the national average (4.4 per cent (engineering), 4.7 per cent (science and agriculture), and 14.7 per cent (health science)) ([13]).

However, this 40–70 per cent figure includes medicine and pharmacy, and needs treating with caution. The staff emphasised that their students' first choice is medical schools at national university. They often try, even if they know that there is a higher chance of failing at their first attempt at the exam.

As obstacles to girls selecting STEM, those mentioned include: 'absence of clear image about scientists' (as opposed to doctors and pharmacists), 'parents' influence', 'lack of role models', 'popular belief that scientists can't make ends meet', and 'lack of hands-on experience making things'. They also recognised the positive effects of studying in an all-girls school on their subject and career choice, although far fewer students now choose to go to a women's university. With varying degrees of agreement across the three schools, many students are mathematically gifted, and that poses no problem to selecting STEM. While some began to apply for US universities in recent years, many are making a very career-driven choice, aiming to secure stability in their life ([12]). These interviews suggest that there is great scope for better communication about the multi-career paths that are now available to women in STEM.

In many countries, initiatives have been taken to start giving awards and grants to female scientists and engineers. For example, a French cosmetic and beauty company L'Oréal is partnered with UNESCO and provides a grant, 'L'Oréal-UNESCO Awards for Women in Science'. Another example in engineering is the STEP (Science, Technology, Engineering and Production) awards in the US, founded in 2013 by the Manufacturing Institute, the University of Phoenix and the Society of Manufacturing Engineers and Deloitte.

Japan has had a series of such prizes and awards (see Appendix A.4). Most recently, Toyota Motor Corporation announced the creation of a fund to support female college and graduate students, studying STEM subjects based in the midland of Japan where the company is located ([14]). The link between higher education and in professional career in industry will become even tighter, and ever closer collaboration among various bodies is inevitable in the future.

As seen in Chapter 4, the original MEXT-funded 'Model' programme was designed to support one institution (either a university or a research institute) only. Since 2013, the new 'hub and spoke' model was introduced to encourage institutional collaboration not only among academic and research institutes, but in also industry. The awarded institution typically set up a liaison committee where partners from all circles (academia, industry and government) contribute.

A little more than one hundred years ago from now, in 1913, only three female students were allowed to study science with their male counterparts at a university in Japan. Their successors including Professor Yuasa, had no choice but to follow and endure, more or less, the same path taken by others, i.e. studying at a women-only university, qualifying as a science teacher and then pursuing their career in research and teaching at the same time. Their individual devotion and sacrifice are beyond anyone's imagination. This is true for all the women in STEM in the world that have persisted in their endeavours.

Well into the twenty-first century, we are in a different place. Now government and industry in Japan are eagerly searching for talent, both male and female, with science and engineering skills and knowledge, and this search is global and borderless. There is no question that a greater number of opportunities are available for Japanese women than previously. However, the hurdles that they come up against may remain still high, accompanying many personal sacrifices, irrespective of formal policy and programmes. In addition, a strong force deriving from the global market and competition may lead to a further fragmentation of collective identity among women in general, and women in STEM.

Women predecessors in STEM have made great strides individually, and we have made great progress collectively in the last century, but there is no room for complacency. Ultimately, the greatest challenge in the future for individuals, families and the nation will be to find a fine balance between tackling the 'why so few' question in science and engineering, while allowing each individual to flourish in society, without stifling the joy of finding their own personal paths and goals in life.

References

[1] T. Tachibanaki. Jyojyo kakusa (Inequality among Women). [In Japanese] Tokyo: Tōyō Keizai Shinpōsha, 2008.

[2] A. Oguni. Hasshin bako: Dansei katsuyaku sokushin hō (opinion: 'Advancement of Men' Bill) [In Japanese]. *Mainichi Shinbun*, 11 November 2014. [online]. Available from: http://mainichi.jp/opinion/news/20141111k0000m070138000c.html. Accessed 17 November 2014.

[3] H. Yoshida. Jyosei katsuyaku sokushin hō' ga mezasu mono towa? (What is the 'Advancement of Women' Bill trying to achieve?) [In Japanese]. Huffington Post Japan, 14 October 2014. [online]. Available from: http://www.huffingtonpost.jp/hiroki-yoshida/woman-activityaction_b_6000858.html. Accessed 17 November 2014.

[4] G. Pascall and J. Lewis. Emerging gender regimes and policies for gender equality in a wider Europe. *Journal of Social Policy*, 33(3): 372–394, 2004.

[5] A. Matysiak. *Interdependencies between Fertility and Women's Labour Supply*, chapter on Women's employment in post-socialist Poland: A barrier or a pre-condition to child-bearing?, pages 111–142. Springer, 2011.

[6] National Tax Agency (NTA). Heisei nijyūyo nenbun minkan kyūyo jittai tōkei chōsa. [In Japanese]. [online]. Available from: https://www.nta.go.jp/kohyo/tokei/kokuzei-cho/minkan/top.htm, 2014. Accessed 12 November 2014.

[7] R.K. Fukuzawa and N. Kodate. *Maternity Services and Policy in an International Context: Risk, Citizenship and Welfare Regimes*, chapter 8: Japan. Abingdon: Routledge, 2015, eds. Patricia Kennedy and Naonori Kodate.

[8] R.K. Fukuzawa and N. Kodate. Maternity services in Japan and Ireland: A comparative perspective. *The University of Tokyo Law and Politics Journal*, 11: 129–158, 2014.

[9] OECD. Growing unequal? income distribution and poverty in OECD countries. country note: Japan. [online]. Available from: http://www.oecd.org/social/soc/41527303.pdf, 2008. Accessed 12 November 2014.

[10] Y. Oka. Daigaku shingaku ritsu no chiikisa, nijyūnen de nibai, daitoshi shūchū de nikyoku ka (Regional differences in college enrollment rates doubled in twenty years: concentration in large metropolitan areas led to bipolarisation). *Asahi Shinbun*, 14 October 2014. http://digital.asahi.com/articles/ASGBG5HCKGBGUTIL03K.html. Accessed 17 November 2014.

[11] Tōyō Keizai Online. Daigaku/gakubu betsu nan-ido rankingu toppu sanjyū (top 30 ranking by university and faculty based on the difficulty to enter) 24 March 2014. [In Japanese]. [online]. Available from: http://toyokeizai.net/articles/-/26876, 2014. Accessed 12 November 2014.

[12] Interviews conducted by Gender Equality Bureau, JST on 14 November 2013, 12 February 2014, 17 February 2014. Personal communications.

[13] Ministry of Education, Culture, Sports, Science and Technology. Gakkō kihon chōsa (School Basic Survey) [In Japanese]. [online]. Available from: http://www.mext.go.jp/b_menu/toukei/chousa01/kihon/kekka/1268046.htm, 2013. Accessed 17 November 2014.

[14] *Nikkei Asia Review*. Toyota hopes to attract female talent through new scholarship. Available from: http://asia.nikkei.com/Business/Companies/Toyota-hopes-to-attract-female-talent-through-new-scholarship, 10 July 2014. Accessed 12 November 2014.

Appendices

A.1 Basic Act for Gender Equal Society (Act No. 78 of 1999)

In consideration of respect for individuals and equality under the law being expressly stipulated in the Constitution of Japan, steady progress has been made in Japan's various efforts toward the realization of genuine equality between men and women, alongside efforts by the international community. However, even greater effort is required.

Further, in order to respond to the rapid changes occurring in Japan's socio-economic situations, such as the trend toward fewer children, the aging of the population, and the maturation of domestic economic activities, it has become a matter of urgent importance to realize Gender Equal society in which men and women respect each other's human rights and share responsibilities, and every citizen is able to fully display their individuality and ability regardless of gender.

In this context, it is vital that the realization of Gender Equal society is positioned as the most important issue in determining the framework of 21st-century Japan and those policies on the promotion of the Formation of Gender Equal Society are implemented in all fields.

This Act is hereby established in order to clarify the basic principles with regard to the Formation of Gender Equal Society, to set a course to this end, and to comprehensively and systematically promote efforts by the State, local governments and citizens with regard to the Formation of Gender Equal Society.

Chapter I General provisions

Article 1 (Purpose)

The purpose of this Act is to comprehensively and systematically promote the Formation of Gender Equal Society, by laying out the basic principles and clarifying the responsibilities of the State, local governments and citizens in regard to the Formation of such a society, and providing fundamental matters for the policies on the promotion of the Formation of Gender Equal Society, in consideration of the urgency of realizing an affluent and dynamic society in which the human rights of both men and women are respected and that can respond to changes in socioeconomic situations.

Article 2 (Definitions)

In this Act, the meanings of the terms listed in the following items shall be as prescribed respectively in those items.

(i) Formation of a Gender-Equal Society: forming a society in which both men and women, as equal members of society, are given opportunities to freely participate in activities in any fields of society and thereby equally enjoy political, economic, social and cultural benefits as well as share responsibilities;
(ii) Positive action: Positive provision of the opportunities stipulated in the preceding item to either men or women to the extent necessary to redress gender disparities in terms of such opportunities.

Article 3 (Respect for the Human Rights of Men and Women)

The Formation of Gender Equal Society shall be carried out, aiming at respect for the dignity of men and women as individuals, not treating men and women in a way that discriminates against them based on gender, securing opportunities for men and women as individuals to exercise their abilities and respect for other human rights of men and women.

Article 4 (Consideration for social systems or practices)

In consideration that social systems or practices can be factors that impede the Formation of Gender Equal Society by reflecting a stereotyped division of roles on the basis of gender, etc., and having a biased effect on men and women's selection of social activities, consideration shall be given so that social systems and practices have as neutral an impact as possible on this selection of social activities.

Article 5 (Joint participation in planning and deciding policies, etc.)

Formation of Gender Equal Society shall be undertaken with the aim to secure opportunities for men and women to jointly participate as equal members of society in planning and deciding policies of the State or local governments or ones of private organizations.

Article 6 (Compatibility of activities in family life and other activities)

Formation of Gender Equal Society shall be undertaken with the aim for men and women who are members of families to be able to smoothly perform their roles as household members with regard to taking care of children and other family members and other activities in their home lives, through mutual cooperation and with social support, and for them to also be able to engage in activities other than these.

Article 7 (International cooperation)

In consideration of close relationship between promotion of the Formation of Gender Equal Society and efforts in the international community, the Formation of Gender Equal Society shall be undertaken based on international cooperation.

Article 8 (Responsibility of the State)

The State is responsible for the comprehensive formulation and implementation of policies on promotion of the Formation of Gender Equal Society (including positive action; the same shall apply hereinafter) pursuant to the basic principles on the Formation of Gender Equal Society prescribed in Articles 3 to the preceding Article (hereinafter referred to as the "Basic Principles").

Article 9 (Responsibility of local governments)

Local governments are responsible for the formulation and implementation of measures that follow the national measures and other measures that reflect the special characteristics of the region of these local governments, in regard to promotion of the Formation of Gender Equal Society, pursuant to the Basic Principles.

Article 10 (Responsibility of citizens)

Citizens shall endeavor to contribute to the Formation of Gender Equal Society in all areas of society, including workplaces, schools, local communities and homes, pursuant to the Basic Principles.

Article 11 (Legislative measures, etc.)

The Government shall take legislative, financial and other measures necessary to implement policies on promotion of the Formation of Gender Equal Society.

Article 12 (Annual reports, etc.)

(1) The Government shall submit an annual report to the Diet on the status of the Formation of Gender Equal Society and the policies implemented by the Government on promotion of the Formation of Gender Equal Society.

(2) The Government shall annually prepare and submit a document to the Diet explaining the policies the Government intends to implement to promote the Formation of Gender Equal Society, considering the status of the Formation of Gender Equal Society described in the report in the preceding paragraph.

Chapter II Basic policies on promotion of the Formation of Gender Equal Society

Article 13 (Basic Plan for Gender Equality)

(1) The Government shall establish a basic plan on the Formation of Gender Equal Society (hereinafter referred to as the "Basic Plan for Gender Equality"), in order to comprehensively and systematically promote policies on the Formation of Gender Equal Society.
(2) The Basic Plan for Gender Equality shall provide for the following matters:

(i) an outline of the policies to be implemented comprehensively and over the long term on promotion of the Formation of Gender Equal Society;
(ii) other than what is listed in the preceding item, matters necessary to comprehensively and systematically implement policies on promotion of the Formation of Gender Equal Society.

(3) The Prime Minister shall hear the opinions of the Council for Gender Equality then formulate a draft of the Basic Plan for Gender Equality and ask the Cabinet for its decision.
(4) The Prime Minister shall publicize the Basic Plan for Gender Equality without delay, when the Cabinet has made its decision pursuant to the provisions of the preceding paragraph.
(5) The provisions of the preceding two paragraphs shall apply mutatis mutandis to changes to the Basic Plan for Gender Equality.

Article 14 (Prefectural Plans for Gender Equality, etc.)

(1) Taking into consideration the Basic Plan for Gender Equality, the prefectures shall establish basic plans with regard to policies on promotion of the Formation of Gender Equal Society with regard to the relevant prefectural areas (hereinafter referred to as "Prefectural Plans for Gender Equality").
(2) Prefectural Plans for Gender Equality shall provide for the following matters:

(i) an outline of the policies to be implemented comprehensively and over the long term on promotion of the Formation of Gender Equal Society regarding the prefectural areas;
(ii) other than what is listed in the preceding item, matters necessary to comprehensively and systematically implement policies on promotion of the Formation of Gender Equal Society within the prefectural area.

3 Taking into consideration the Basic Plan for Gender Equality and Prefectural Plans for Gender Equality, the municipalities shall endeavor to establish basic plans with regard to policies on promotion of the Formation of Gender Equal Society within the relevant municipal areas (hereinafter referred to as "Municipal Plans for Gender Equality").
4 When a prefecture or a municipality establishes or modifies its Prefectural Plan for Gender Equality or its Municipal Plan for Gender Equality, the prefecture or municipality shall publicize this without delay.

Article 15 (Consideration in the formulation of policies, etc.)

When formulating and implementing policies that are found to influence the Formation of Gender Equal Society, the State and local governments shall give consideration to the Formation of Gender Equal Society.

Article 16 (Measures to increase the understanding of the citizens)

The State and local governments shall take appropriate measures through public relations activities, etc., to increase citizens' understanding of the Basic Principles.

Article 17 (Processing of complaints, etc.)

The State shall take necessary measures for processing complaints in regard to policies implemented by the Government which are related to promotion of the Formation of Gender Equal Society or which are found to influence the Formation of Gender Equal Society, and necessary measures intended to provide relief to victims whose human rights have been infringed through gender-based discriminatory treatment and other factors impeding the Formation of Gender Equal Society.

Article 18 (Research and study)

The State shall endeavor to promote necessary research and study into the formulation of policies concerning the promotion of the Formation of Gender Equal Society, including research and study related to the impact of social systems and customary practices on the Formation of Gender Equal Society.

Article 19 (Measures for international cooperation)

In order to promote the Formation of Gender Equal Society based on international cooperation, the State shall endeavor to take necessary measures for the exchange information with foreign governments and international organizations, and for the smooth promotion of international mutual cooperation related to the Formation of Gender Equal Society.

Article 20 (Support for local governments and private organizations)

The State shall endeavor to provide information and take other necessary measures in order to support policies on promotion of the Formation of Gender Equal Society implemented by local governments, and activities on promotion of the Formation of Gender Equal Society undertaken by private organizations.

Chapter III Council for Gender Equality

Article 21 (Establishment)

The Council for Gender Equality (hereinafter referred to as the "Council") is hereby established in the Cabinet Office.

Article 22 (Affairs under Council jurisdiction)

The Council is in charge of the following administrative affairs:

(i) dealing with the matters provided in paragraph (3) of Article 13 with regard to the Basic Plan for Gender Equality;
(ii) in addition to what is listed in the preceding item, studying and deliberating on basic principles, basic policies and important matters with regard to promotion of the Formation of Gender Equal Society in response to consultations with the Prime Minister or relevant ministers;
(iii) studying and deliberating on matters provided in the preceding two items, and, if necessary, stating opinions to the Prime Minister and relevant ministers ; and
(iv) monitoring the status of implementation of government measures to promote the Formation of Gender Equal Society, and studying the impact of government measures on the Formation of Gender Equal Society, and, if necessary, stating opinions to the Prime Minister and relevant ministers.

Article 23 (Organization)

The Council shall consist of a chairperson and no more than 24 members.

Article 24 (Chairperson)

(1) The Chief Cabinet Secretary shall serve as chairperson.
(2) Chairperson shall preside over the affairs of the Council.

Article 25(Council Members)

(1) The persons listed in following items shall serve as Council members:

(i) those who are designated by the Prime Minister, among Cabinet ministers except for the Chief Cabinet Secretary;

(ii) persons appointed by the Prime Minister among those who have distinguished insight into the Formation of Gender Equal Society.

(2) The number of members in item (ii) of the preceding paragraph shall not be less than half of the number of members provided for in the same paragraph.

(3) Neither the number of male nor the number of female members in item (ii) of paragraph (1) may fall below four-tenths of the total number of the members.

(4) Members prescribed in item (ii) of paragraph (1) shall work on a part-time basis.

Article 26 (Term of Members)

(1) Members prescribed in item (ii) of paragraph (1) of Article 25 shall serve a term of two years. A member who is appointed to fill a vacancy shall serve the remainder of the predecessor's term.

(2) Members prescribed in item (ii) of paragraph (1) of the preceding Article may be reappointed.

Article 27(Request for Submitting Documents, etc.)

(1) The Council may request the heads of relevant administrative organs to submit necessary materials for monitoring or investigation and any other materials, state an opinion, give an explanation, or give any other necessary cooperation, when it finds it necessary for pursuing affairs under its jurisdiction.

(2) When the Council finds it especially necessary for pursuing affairs under its jurisdiction, the Council may also request the necessary cooperation from a person other than one provided for in the preceding paragraph.

Article 28 (Delegation to Cabinet Order)

In addition to what is provided for in this Chapter, the organization of the Council, members and other officials, and other necessary matters on the Council shall be prescribed by Cabinet Order.

Supplementary Provisions [Act No. 160 of December 22, 1999] [Extract]

Article 1(Effective Date)

This Act (excluding Article 2 and 3) shall come into force as of January 6, 2001.

A.2 The Basic Plan for Gender Equality

I Building social systems that promote gender equality

1. Increasing participation by women in policy decision-making processes

(1) Increasing participation by women in national policy decision-making processes

   (a) Promoting participation by women in national advisory councils and committees
   (b) Promoting recruitment and appointment of women national civil servants

(2) Request for the support and cooperation of local governments

   (a) Support of efforts related to participation by women in advisory councils and committees
   (b) Request cooperation related to the recruitment and appointment of women civil servants in local governments

(3) Support for efforts by private companies, educational/research institutes and other types of organizations
(4) Conducting inquiries and collecting/providing information/data

   (a) Conducting inquiries/research on participation in policy decision-making
   (b) Collection, organization and provision of information on female human resources

2. Reviewing social systems and practices and reforming awareness from the perspective of gender equality

(1) Review of social systems and practices from the perspective of gender equality
(2) Carrying out informative and educational activities to reach out to a broad sector of the public
(3) Strengthening legal literacy
(4) Collection, organization and provision of information related to gender equality

II Achieving gender equality in the workplace, family and community

3. Assuring equal opportunity and equal treatment in employment

(1) Promoting equal opportunity and equal treatment in employment
(2) Protection of working women during pregnancy/after childbirth
(3) Developing the work talents of women and supporting the display of those talents
(4) Establishing employment conditions that enable various working patterns

   (a) Comprehensive promotion of part-time work measures
   (b) Promoting measures related to worker dispatching enterprises
   (c) Supporting women entrepreneurs
   (d) Support for family employees
   (e) Promotion of measures concerning new types of work

4. Building partnerships in agriculture, forestry, and fishing villages

(1) Reforming awareness and actions in all fora
(2) Increasing participation by women in policy decision-making processes

(3) Improving the economic status of women and improving work conditions/ environment
(4) Creating an environment conducive to the lives and activities of women
(5) Establishing conditions where the elderly can lead active, secure lives

5. *Supporting women and men to harmonize work with family and community life*

(1) Enhancing measures to support the alignment of child-raising with diverse ways of life

  (a) Enhancing measures to support child-raising aligned with diverse ways of life
  (b) Enhancement of support for single-parent families

(2) Improving the environment for the continued employment of workers engaged in child care/family care

  (a) Establishing an environment that makes it easy to take child care/family care leave and return to the workplace
  (b) Establishing an environment that makes it easy to continue work while engaging in child care and family care

(3) Assistance for those seeking re-employment

  (a) Promoting extension of re-employment systems
  (b) Support for those seeking re-employment

(4) Promoting joint participation between women and men in family life and community life

  (a) Promoting joint participation between women and men in family life
  (b) Promoting joint participation between women and men in the community
  (c) Improving work conditions (e.g. shortening work hours)

6. *Developing conditions to enable elderly people and others to live with peace of mind*

(1) Building a family care system that enables the elderly to live with a peace of mind

  (a) Promoting health and welfare measures for the elderly
  (b) Securing nursing personnel
  (c) Establishing a new nursing system
  (d) Establishing social infrastructure conducive to encouraging the independence of the elderly

(2) Enhancing income guarantees for the elderly
(3) Promoting the participation of the elderly in society
(4) Priority on the consideration of those with disabilities

III Creating a society where the human rights of women are promoted and defended

7. Elimination of all forms of violence against women

(1) Dealing strictly with violence against women

- (a) Dealing strictly with sexual crimes
- (b) Taking comprehensive measures against prostitution
- (c) Ascertaining conditions and promoting countermeasures against domestic violence and other such violence which tends to go unreported
- (d) Promoting measures to prevent sexual harassment

(2) Enhancement of recuperative measures for women victims of violence

- (a) Enhancing consultation, protection and relief measures for women victims
- (b) Training personnel involved in interviewing victims, prosecution and providing advice, relief, etc., in relation to incidents of violence against women

(3) Creating an environment for the prevention of violence against women

- (a) Publicity, education and elimination of harmful environments toward a society that prohibits violence against women
- (b) Strengthening crime prevention
- (c) Studies and research on the causes and impact of crime and related prevention measures

(4) Strengthening cooperation with concerned organizations toward the elimination of violence against women, and studying comprehensive measures

8. Respect for human rights of women in the media

(1) Support for media efforts to adopt expressions that respect the human rights of women

- (a) Promoting respect for human rights in the media and measures to keep expressions of sex and violence away from those who do not wish to come into contact with them
- (b) Considering the establishment of rules for the Internet and other new media
- (c) Promoting gender-free expressions in official releases and publications of official organizations

9. Supporting life-long health for women

(1) Spreading awareness of reproductive health/rights

(2) Advancing measures to maintain and promote women's health throughout the life cycle

- (a) Enhancing health education and counseling support to maintain and promote health throughout the life cycle

(b) Supporting women's health during pregnancy and childbirth
(c) Support for the promotion of women's health during adulthood and old age

(3) Promoting measures to address problems that threaten women's health

(a) Countermeasures against HIV/AIDS, sexually transmitted diseases
(b) Promoting measures to prevent drug abuse

10. Promoting gender equality to open the way for a diversity of choice through enrichment of education and learning

(1) Education and learning to promote gender equality

(a) Enhancing primary and secondary education
(b) Promoting gender equality in institutions of higher learning
(c) Promoting social education

(2) Enhancing educational and learning opportunities to enable various choices

(a) Promoting lifelong learning
(b) Enhancing education and learning activities matched to the diversified and sophisticated learning demands of women
(c) Enhancing career/employment guidance

IV Contributing to the 'equality, development and peace' of the global community

(1) Absorbing international norms and standards into Japan
(2) Contributing to the 'equality, development and peace' of the global community

(a) Cooperation in the activities of the United Nations
(b) Promotion of Women in Development (WID)
(c) Contributions of women to peace
(d) Promoting participation by women in the policy decision-making process in international arenas
(e) Promoting international exchange and cooperation

(Source: Gender Equality Bureau website, http://www.gender.go.jp/english_contents/about_danjo/lbp/basic/koudou/contents.html). Accessed 20 November 2014.

A.3 List of 34 female ministers in Japan, 1960–2013

Prime Minister	*Name of final degree awarding institution*	*Ministerial post*	*Period*	*Party*
Hayato Ikeda	Masa Nakayama (Ohio Wesleyan University)	Health and Welfare	19 July 1960–8 December 1960	LDP
	Tsuruyo Kondō (JWU)	Science and Technology Agency	18 July 1962– 18 July 1963	LDP
Yasuhiro Nakasone	Shigeru Ishimoto (Red Cross Nursing School)	Environment Agency	1 November 1984–28 December 1985	LDP
Toshiki Kaifu	Mayumi Maruyama (Tsuda/ University of Tokyo)	Environment Agency	10 August 1989 – 25 August 1989	LDP
		Chief Cabinet Secretary	15 August 1989 – 28 February 1990	
	Sumiko Takahara (Hitotsubashi)	Economic Planning Agency	10 August 1989 – 28 February 1990	Non-partisan
	Akiko Santō	Science and Technology Agency	29 December 1990 – 5 November 1991	LDP
Kiichi Miyazawa	Mayumi Moriyama	Education	12 December 1992 – 9 August 1993	LDP
Morihiro Hosokawa	Ryōko Akamatsu (University of Tokyo)	Education	9 August 1993 – 28 April 1994	Non-partisan
	Sanae Kubota (Tsuda/Keiō University)	Economic Planning Agency	9 August 1993 – 28 April 1994	Japan Socialist Party
	Wakako Hironaka (Ochanimizu/ Brandeis University)	Environment Agency	9 August 1993 – 28 April 1994	Kōmei
Tsutomu Hata	Ryōko Akamatsu	Education	28 April 1994 –30 June 1994	Non-partisan
	Toshiko Hamayotsu (Keiō)	Environment Agency	28 April 1994 – 30 June 1994	Kōmei
Tomiichi Murayama	Makiko Tanaka (Waseda University)	Science and Technology Agency	30 June 1994 – 8 August 1995	LDP
Ryūtarō Hashimoto	Ritsuko Nagao (University of Tokyo)	Justice	11 January 1996 – 7 November 1996	Non-partisan
	Michiko Ishii (Tokyo University of Pharmacy and Life Sciences)	Environment Agency	7 November 1996–11 September 1997	LDP
Keizō Obuchi	Seiko Noda (Sophia University)	Posts and Telecommunications	29 July 1998 – 5 October 1999	LDP
	Kayoko Shimizu (University of Tokyo)	Environment Agency	5 October 1999 – 5 April 2000	LDP

(Continued)

Prime Minister	*Name of final degree awarding institution*	*Ministerial post*	*Period*	*Party*
Yoshiro Mori	Kayoko Shimizu	Environment Agency	5 April 2000 – 4 July 2000	LDP
	Chikage Ōgi	Construction	4 July 2000 – 5 January 2001	Conservative
		Transportation/Hokkaidō Development Agency/ National Land Agency	5 December 2000 – 5 January 2001	
		Land, Infrastructure and Transportation	6 January 2001 – 22 September 2003	
	Yoriko Kawaguchi (University of Tokyo/Yale University)	Environment Agency	4 July 2000 – 6 January 2001	Non-partisan
		Environment	6 January 2001 – 26 April 2001	
Junichirō Koizumi	Mayumi Moriyama	Justice	26 April 2001 – 22 September 2003	LDP
	Makiko Tanaka	Foreign Affairs	26 April 2001 – 30 January 2002	LDP
	Yoriko Kawaguchi	Environment	26 April 2001 – 30 September 2002	Non-partisan
		Foreign Affairs	1 February 2002 – 27 September 2004	
	Atsuko Tōyama (University of Tokyo)	Education, Culture, Sports, Science and Technology	26 April 2001 – 22 September 2003	Non-partisan
	Yuriko Koike (Cairo University)	Environment	19 November 2003 – 26 September 2006	LDP
		Okinawa and Northern Territories Affairs	27 September 2004 – 26 September 2006	
	Kiyoko Ono (Tokyo University of Education)	National Public Security Commission/Youth Affairs and Measures for Declining Birthrate/Food Security	22 September 2003 – 27 September 2004	LDP
	Chieko Nōno (Osaka University)	Justice/Youth Affairs and Measures for Declining Birthrate	27 September 2004 – 30 October 2005	LDP
	Kuniko Inoguchi (Sophia University/Yale University)	Measures for Declining Birthrate and Gender Equality	30 October 2005 – 26 September 2006	LDP

Shinzō Abe (1st)	Hiroko Ōta (Hitotsubashi)	Economic and Fiscal Policy	26 September 2006 – 25 September 2007	Non-partisan
	Sanae Takaichi (Kobe University)	Okinawa and Northern Territories Affairs/Science and Technology Policy/Innovation/ Measures for Declining Birthrate and Gender Equality/ Food Security	26 September 2006 – 27 August 2007	LDP
	Yuriko Koike	Defense	4 July 2007 – 27– August 2007	LDP
	Yoko Kamikawa (University of Tokyo)	Measures for Declining Birthrate and Gender Equality	27 August 2007 – 25 September 2007	LDP
Yasuo Fukuda	Hiroko Ōta	Economic and Fiscal Policy	26 September 2007 – 2 August 2008	Non-partisan
	Yōko Kamikawa	Measures for Declining Birthrate and Gender Equality	26 September 2007 – 2 August 2008	LDP
	Seiko Noda	Consumer Agency	2 August 2008 – 24 September 2008	LDP
	Kyōko Nakayama (University of Tokyo)	Abduction Issue	2 August 2008 – 24 September 2008	LDP
Tarō Asō	Seiko Noda	Consumer Agency	24 September 2008 – 16 September 2009	LDP
		Science and Technology/Food Security	24 September 2008 – 16 September 2009	
	Yūko Obuchi (Seijō University/ Waseda University)	Measures for Declining Birthrate and Gender Equality	24 September 2008 – 16 September 2009	LDP
Yukio Hatoyama	Keiko Chiba (Chūō University)	Justice	16 September 2009 – 8 June 2010	Democratic Party of Japan (DPJ)
	Mizuho Fukushima (University of Tokyo)	Measures for Declining Birthrate and Gender Equality	16 September 2009 – 28 May 2010	Social Democratic Party

(Continued)

Prime Minister	*Name of final degree awarding institution*	*Ministerial post*	*Period*	*Party*
Naoto Kan	Keiko Chiba	Justice	8 June 2010 – 17 September 2010	DPJ
	Renhō Murata (Aoyama Gakuin)	Administrative Reform	8 June 2010 – 27 June 2011	DPJ
	Tomiko Okazaki (Fukushima Prefecture Girls High School)	National Public Security Commission	17 September 2010 – 14 January 2011	DPJ
Yoshihiko Noda	Renhō Murata	Administrative Reform	2 September 2011 –13 January 2012	DPJ
	Yōko Komiyama (Seijō University)	Health, Labour and Welfare	2 September 2011 –1 October 2012	DPJ
	Makiko Tanaka	Education, Culture, Sports, Science and Technology	1 October 2013 –26 December 2012	DPJ
Shinzō Abe (2nd)	Tomomi Inada (Waseda University)	Regulatory Reform	26 December 2012 – 3 September 2014	LDP
	Masako Mori (Tōhoku University/ New York University)	Measures for Declining Birthrate and Gender Equality	26 December 2012 – 3 September 2014	LDP

(*Source*: Data collated by the authors.)

A.4 A list of awards and research grants only available to Japanese female researchers in STEM (excluding bursaries for students)

Award/Grant	*Awarding body*	*Amount of grant, etc.*
Banyu Foundation Research Grant 2014 (Support for Woman Scientist)	Banyu Life Science Foundation International	2,000,000 JPY (1,000,000 JPY per year for 2 years)
Female Researcher Encouragement Award (subsidy for 'childcare for building a stronger future')	Sumitomo Life Insurance Company	Up to 1,000,000 JPY per year for 2 years
Hokkaidō University Ōtsuka Award	Hokkaidō University	500,000 JPY (for seven postgraduate students per year, who seek to become a researcher)
Inoue Science Research Award	Inoue Foundation for Science	5,000,000 JPY (out of four winners, at least one should be female)
Irisawa Aya Memorial Female Physiologist Encouragement Award (Irisawa Aya Award)	Physiological Society of Japan	Certificate of merit, supplementary prize (1,000,000 JPY)
JWEF Encouragement Award for Female Engineers	Japan Women Engineer Forum	Commemorative shield, supplementary prize (100,000 JPY)
Woman Researcher Award (Nakamura Award)	Kanazawa University	Certificate of merit, supplementary prize (100,000 JPY)
Woman Researcher Encouragement Award (Nakamura Award)	Kanazawa University	Certificate of merit, supplementary prize (50,000 JPY)
The Most Outstanding Woman Scientist Award (Tachibana Award)	Kyoto University	Certificate of merit, memorabilia, supplementary prize (100,000 JPY)
Outstanding Woman Scientist Encouragement Award	Kyoto University	Certificate of merit, memorabilia
L'Oréal-UNESCO Award For Women in Science (Japan)	L'Oréal Japan	1,000,000 JPY
Morita Award for Scientific Research	Japanese Association of University Women	Certificate of merit, supplementary prize (500,000 JPY)

(Continued)

Award/Grant	*Awarding body*	*Amount of grant, etc.*
Outstanding Female Researcher Award & Contributions to Investing in People Award (Kodate Kashiko Award)	Japan Society of Applied Physics	Certificate of merit, memorabilia
Saruhashi Award	The Society for Bringing Forth a Bright Future for Female Scientists	Certificate of merit, supplementary prize (300,000 JPY)
Shiseidō Female Research Science Grant	Shiseidō	1,000,000 JPY
SJWS Woman Scientist Encouragement Award	The Society of Japanese Women Scientists	Certificate of merit, encouragement prize (200,000 JPY)
SJWS Woman Scientist Achievement Award	The Society of Japanese Women Scientists	Certificate of merit, memorabilia
Subsidy for Female Researchers after Maternity Leave	The Naito Foundation	2,000,000 JPY per year for 3 years
Tōhuku University Gender Equality Promotion (Sawayanagi) Award	Tōhuku University	3 categories (A: research, B: activity and C: project). Certificate of merit, research grant (1,400,000 JPY in total, covering 3 awards)
Woman Chemist Award	Chemical Society of Japan	Commemorative shield
Woman Researcher Encouragement OM Award	The Zoological Society of Japan	500,000 JPY

(Data collated by the authors. These exclude awards for female undergraduate students.)

Index

For Product Safety Concerns and Information please contact our EU representative GPSR@taylorandfrancis.com
Taylor & Francis Verlag GmbH, Kaufingerstraße 24, 80331 München, Germany

www.ingramcontent.com/pod-product-compliance
Ingram Content Group UK Ltd.
Pitfield, Milton Keynes, MK11 3LW, UK
UKHW022124190926
475702UK00013B/1641
9780815383901